From Being to Doing

Humberto R. Maturana/Bernhard Poerksen
The Origins of the Biology of Cognition

Translated by Wolfram Karl Koeck and Alison Rosemary Koeck

2004

Published by Carl-Auer Verlag: **www.carl-auer.com**
Please order our catalogue:

Carl-Auer Verlag
Weberstrasse 2
69120 Heidelberg
Germany

Layout: Verlagsservice Josef Hegele, Dossenheim
Cover: WSP Design, Heidelberg
Coverpicture: Rene Magritte „The False Mirror"
© Bild-Kunst, Bonn 2004
Printed in The Netherlands
Printed by Koninklijke Wöhrmann B. V., Zutphen

ISBN 3-89670-448-6

Copyright © 2004 by Carl-Auer-Systeme
All rights reserved. No part of this book may be reproduced by any process whatsoever without the written permission of the copyright owner.

Title of the original edition:
„Vom Sein zum Tun"
© 2002 by Carl-Auer-Systeme, Heidelberg

Bibliographic information published by Die Deutschen Bibliothek
Die Deutsche Bibliothek lists this publication
in the Deutsche Nationalbibliografie; detailed bibliographic
data is available in the Internet at http://dnb.ddb.de.

Contents

Preface ... 9
Acknowledgments ... 10
Introduction for the English edition ... 12
Introduction ... 16

I. The cosmos, an explanation of observing ... 25

1. Without the observer, there is nothing ... 26
Everything that can be said is said ... 26
In the beginning was the difference ... 29
The explanation of experience ... 32
The age of self-observation ... 34

2. Varieties of objectivity ... 38
Life in the multiverse ... 38
A multitude of worlds ... 42
Tolerance and respect ... 46
Aesthetic seduction ... 50

3. The biology of cognition 53
The experience of truth ... 53
The epistemology of an experiment ... 54
Why the nervous system is closed ... 58
Double look ... 62
To live is to know ... 66

4. On the autonomy of systems ... 68
The limits of external determination ... 68
Organisation and structure ... 71
Understanding responsibility ... 75
A miracle is needed ... 79

5. How closed systems interact ... 82
Improbable interactions ... 82
Structural coupling ... 85
The myth of successful communication ... 88
The world arises in language ... 90

6. Autopoiesis is living ... 93
Confrontation with death ... 93
A factory that produces itself ... 96
Autopoietic and allopoietic systems ... 100
The second creation ... 102

7. The history of an idea ... 104
A concept becomes fashionable ... 104
Imploring Erich Jantsch on bended knees ... 105
Human beings are indispensable ... 108
Systems theory as worldview ... 109

II. Application of a theory ... 113
1. Psychotherapy ... 114
The view of the systemicist ... 114
Change of change ... 117
Individual and society ... 121
The constructions of pathology ... 122

2. Education ... 127
The paradox of education ... 127
Listening to the listening ... 129
Perception and illusion ... 132
All human beings are equally intelligent ... 135

III. History of a theory ... 139
1. Beginnings and inspirations ... 140
Insights of a child ... 140
The warm-blooded dinosaur ... 143
What the frog's eye tells the frog's brain ... 146

2. Return to Chile ... 150
Competition means dependency ... 150
Insights of an outsider ... 153
The *Tractatus biologico-philosophicus* ... 157
Systemic wisdom ... 161
The brain of a country ... 165

3. Experience of a dictatorship ... 167
The emergence of blind spots ... 167
Ideology and the military ... 169
The powerlessness of power ... 171
The maintenance of self-respect ... 175
Encounter with Pinochet ... 178

4. Worlds of science ... 183
The paradogma ... 183
Between philosophy and science ... 185
Notes of an observer ... 188
The doors of perception ... 190

IV. Ethics of a theory ... 195
1. The biology of love ... 196
The two identities of the scientist ... 196
Trusting existence ... 198
Social systems ... 202
Ethics without morality ... 206

Preface

Humberto Maturana, whom I have known for nearly half a century, always addresses his audiences, whether philosophers, physicists, family therapists, business executives or others, with the words: "Whoever I am talking to, I'm talking to as a biologist." He maintains this attitude in the fascinating conversations with Bernhard Poerksen, a perceptive and intelligent partner, which have resulted in an impressive panorama of ideas stretching from the intricate problems of philosophy and logic to the fundamental ethical questions of everyday life. The central point of view chosen here is the point of view of life itself. Wherever one opens this rewarding book, one will close it again with an enriched and stimulated mind.

Heinz von Foerster
Prof. h. c. University of Vienna, Prof. em. University of Illinois,
Rattlesnake Hill, February 2002

Acknowledgments

Humberto R. Maturana and I met for the first time in May 2000 in his rooms at the University of Chile in the centre of Santiago. It was there, in his laboratory, that the plan took shape to compile a book that would present, in dialogical form, Maturana's *neurosophy*, that special mixture of rigorous and wild thinking along the borderlines of natural science and philosophy. During this first encounter we reached some agreement about the key topics and talked, still quite warily and hesitantly and groping for the right kind of form, about the discovery of the observer and the biology of cognition. Torrential downpours, however, flooded half of Santiago so badly that one could only move around in rubber dinghies, and so we could not see each other often enough. The definitive meetings that finally produced this book took place in March 2001, again in Santiago. Our discussions and debates, which varied widely in content, always revolved around a decisive transformation, a re-orientation *from being to doing*, from the essence of an object to the process of its production. And whatever the topic – the era of the dictatorship in Chile, the education of children, or the theory of autopoiesis –, Humberto R. Maturana invariably focusses on foundational issues, full of enthusiasm but with intellectual rigour. It is the conditions that generate a reality, that bring it forth, in the first place, that fascinate him, and that he seeks to explore. From such a perspective, nothing remains unchangeable and simply given, everything may be related to and explained by its particular origin and development. When writing this book, I tried very hard to preserve as much as possible of the spirit and the dynamics of this kind of thinking fascinated by changes and transformations. The publisher, Carl-Auer-Systeme, Heidelberg, has been most helpful. Ralf Holtzmann and Klaus W. Mueller have supported the project with confidence and stimulating optimism. Wolfram K. Koeck, who translated the introduction into German and helped with my adaptation, was always available when problems arose with the German version.[1] Matthias Eckoldt, Julia Raabe and Friederike Stock looked through the first transcriptions and formulated their critical comments in such a charming manner that they became inspirations. But the book would never have seen the light of day in its present form without Humberto R. Maturana himself and his practically inexhaustible willingness to talk to me. It could not have been written without his dedication and trust. He therefore deserves my very special, heartfelt gratitude.

Bernhard Poerksen
Hamburg, April 2004

---

[1] The English translation of the book was prepared by Wolfram Karl Koeck and Alison Rosemary Koeck. The present version contains original English contributions by Humberto R. Maturana ("Introduction"; texts accompanying figs. 1–12) and occasional rewordings by the authors and the publisher.

Introduction for the English edition

This book presents a rather long conversation that I had with Bernhard Poerksen about the history of my work on the biology of cognition. It is no more but no less than that. So I have not much more to say in this short preface than what I have already said in the book. Yet, I would like to add some reflections on how I lived what the book tells. In particular I will reflect on three basic turning points that I lived while I was working in what became the biology of cognition and the biology of love.

The three turning points that I am talking about occurred to me in relation to my becoming aware of the systemic implications of three ordinary features of our daily living. They were the relational nature of questions, the ordinary fact that we commit mistakes, and our normal daily trust in the repetitiveness of natural phenomena. Of course I knew that questions take place in the relation of the person that asks the question and the person that answers it. Of course I knew that I committed mistakes, and of course I knew that I trusted the regularity of natural processes in my daily living. The expansion in my awareness referred to my becoming conscious of the consequences of acting in the awareness of what those ordinary circumstances and processes of our daily living entail for our doings and our understanding of what we do. Let us see:

Questions and answers

If we attend to the relational nature of questions and answers, we can easily see that the person that accepts an answer to his or her question determines in his or her listening what makes the answer that he or she accepts valid for him or her. Whatever the question may be, it is a constitutive feature of the question answer relation that the person that accepts the answer determines what makes it a valid answer. Yet, this is not a peculiar feature of questions and answers; in every relation in which something offered is accepted, the person that accepts what is offered determines the truth, value, or adequacy of what is accepted. Of course what I say is not new, indeed is well known. Yet, if we accept that that is indeed the case, we cannot henceforth ignore in what we do that nothing is true in itself, valuable, adequate or acceptable in itself. Furthermore, if we accept the implications of what I have said above, the following questions arise: what is to know? What is the sense of fighting for the truth? When a scientist asks a question to nature and obtains an answer through experiments or observations, is he or she aware of the fact that it is he or she who determines the validity of the answer obtained, by choosing the criterion that he or she uses to accept or to reject the results of the experiments or the observations?

When I became aware of the fact that it is the observer who decides the validity of what he or she accepts as valid, and that that is a constitutive feature of the relation question and answer, I realised that the questions proposed above had to be answered taking that into consideration.

We commit mistakes

We live as if we had in some way a direct or an indirect access to that which we call reality to validate our statements or explanations. Yet, we commit mistakes. We say that we learn through our mistakes, but we punish others, whoever they may be, politicians, children, scientists, parents, philosophers... for the mistakes that they commit. What does this reveal? We treat mistakes as serious failures in our behaviour that reveal a guilty blindness in front of a reality that we should see because we have the ability to do so.

If we ask ourselves what occurs when a mistake is committed, we shall easily see that a mistake is an action done in the honest acceptance of its validity in the moment that it is done, and that is later devaluated as a mistake in relation to an other action whose validity is accepted without doubt. But, to the extent that this is so, mistakes are not mistakes in themselves, they are not failures, they do not reveal our blindness about reality. Mistakes do not happen in the moment in which we say that they occurred, they happen afterwards when we compare actions occurring in successive moments. We do not know that we commit a mistake when we commit a mistake. Mistakes do not occur in the present, they occur afterwards. If we had know that what we were doing was not valid in the moment of doing it, we would have been lying. Mistakes are not faults, mistakes are not failures of our capacities, mistakes do not show our limitations, mistakes arise as reflections on the course of our doings. But, if we do not know in the moment in which we do whatever we do, whether we shall later see that doing as a mistake in relation to something else which we do not know either if we shall later see this other doing as a mistake, in what sense could we claim to have access to an independent reality to validate what we do? In what sense can I claim that I know the truth, or how things are, if I do not know if I shall later think that such claim was a mistake? Why should any one be punished for committing a mistake? What is to know, then?

When I became conscious of the fact that mistakes are not in themselves, that they do not occur in the present, and that they occur after the action that is later called a mistake has been done, arising in a posterior act of reflection, I thought that the question "what is to know?" had to be answered accepting that we never know in the moment that we do what we do if we shall later call it a mistake.

Trusting the repetitiveness of nature

We move in daily life trusting that that which we call nature is repetitive, trusting that that which worked once will work again if the proper conditions are realised. This trust is the fundament of all that we do in our daily living, whatever this may be, cooking, gardening, science, technology or philosophy. This, of course we all know. Moreover, we all know that the things that we make, as well as those that are natural, operate according to the way they are made, and we trust that. Of this we are probably all aware as we operate in our daily life. But of what we are not all aware is of the fact that to the extent that natural and artificial "things" operate according to how they are made, we cannot specify by acting on them what happens to them, and all that we can do is to trigger in them changes that arise determined by the manner they are made. We as living systems are not an exemption, as molecular entities we are like all other molecular entities, and what happens to us at any instant is determined in us by the way we are made at that instant, and not by the external agents that impinge upon us.

When I became conscious of the fact that external agents do not specify what happens in us, and that they only trigger in us changes determined by the way we are made, I asked myself, what is to know then? How will anything external to me tell me anything about itself if what I see, hear or accept, is determined by the way I am made? In these circumstances the question, what is to know? has to be answered accepting as part of our natural existence the fact that nothing external to us can tell us anything about itself.

As I became progressively aware of the broad implications of these features of our daily living, my understanding of biological processes expanded and began to change. I began to be aware of the processes that gave origin to whatever I distinguished, and instead of asking about how things were, I began asking for the processes that gave origin to them, and for the criteria that I used to accept the answers that I considered valid. This book is thus the history of a change of question, the history of going from the question how is that?, to the question, what criterion do I use to claim that something is as I say that it is?

Reflections

In this preface I am doing a philosophical reflection about my work because I am reflecting on the fundaments of what I say, not because I am a professional philosopher, which I am not. All human beings do philosophical reflections when they ask about the fundaments of their beliefs or of what they think they know. I also think that one does science whenever one proposes a process that would generate, as a consequence of its operation, some experience that one wants to explain. This book is also the history of some philosophical reflections and of the scientific answers to which the questions that arose from those reflections.

As such in this book I tell my life, and I thank the reader for making me the gift of reading it.

Humberto R. Maturana
Santiago de Chile, April 2004

Introduction

Human life occurs in daily living. This statement sounds obvious, and it is so. Yet, by saying it I want to emphasise that all our activities, regardless of whether they are homely, artistic, professional, or technical, are only particular cases of our daily living, and *do not* entail anything different from what we do in our home chores other than the special features of the relational and operational spaces in which they take place, or the different purposes, aims or desires under which we do what we do. This book is a reflection about how we do whatever we do, and about the history of how the various notions presented in it arose in the course of my daily living in the attempt of understanding how we see, how we hear, … and in general how we know what we claim to know.

I was an ordinary child with an ordinary living, and the only thing that perhaps was in some way peculiar in me was that I have conserved as features of my daily concerns certain questions that arose in me as a child. And as I conserved these questions I lived them as if they were aspects of my daily living that I wanted to answer with the elements of my daily living. This was not trivial. Somehow I was not interested in essences. I did not want to know how things were in themselves. I wanted to know how they happened. I loved to make my own toys, I loved to climb trees and to listen to the many sounds that the insects made. I loved insects, crabs, plants, animals in general, and I liked to collect the hard remains of their bodies, to see how they related to each other and to their manner of living.

I liked to move, to jump, to walk and to run, and in that way I knew my body as well as the different worlds in which I existed as they arose with my movements and live them in the pleasure of doing whatever I did. I felt that I was like the insects and the crabs that I liked to contemplate, and whose skeletons I liked to examine to see how they moved in relation to the way they lived. I lived in doing. I saw in doing. I thought in doings. This just happened to me. Yet as a child of my culture I lived at the same time in a world that happened around me and existed outside of me by itself.

This book reveals the history of a metaphysical change in my thinking, in my feelings and my way of understanding life and the worlds I live. This book does not contain the history of the reflections of a philosopher or the history of the doings of a scientist, it contains the history of some aspects of the experimental research and philosophical reflections of a biologist interested in understanding living, perception, and cognition as a feature of the continuous flow of the living of living systems in general, and of us human beings in particular. Therefore, although this book does not contain the history of a scientific quest, it tells of the history of the expansion of the understanding of life and of humanness that takes place when a biologist accepts as a matter of daily experience that all that living systems in general, and all that human beings in particular, do and experience takes place in the realisation of their living as living systems, and thinks that life, cognition, and consciousness are biological phenomena to be explained as such with the features of the coherences of living without additional assumptions.

Our present patriarchal-matriarchal culture is lived in an implicit, and sometimes explicit metaphysical view that entails accepting as a matter of course that existence occurs in a background of essences that exist independently of what we human beings do. I call this metaphysical attitude or fundamental reflective standing point of our patriarchal-matriarchal culture *the metaphysics of the transcendental reality.*

Our patriarchal-matriarchal culture is centred around the separation of what is apparent from what is essential under the spell of the question that asks for what is, for what is real, rather than for what do we do when we claim that something is the case. In this culture we live in the search of our essential being, our true self, in a quest that proves again and again impossible to fulfil because at the same time we accept *a priori* that that question does not have an answer in the domain of our daily living which is where in fact we do all that we do. And , as a result, we are forced to fall again and again either into total scepticism about our possibility of understanding ourselves as selfconscious languaging systems, or we are forced to fall in a sort of theological thinking to justify our biologically unexplainable existence as human beings.

This book shows how I abandoned the metaphysical attitude of our culture that takes for granted the existence of an independent reality as the transcendental background on which everything occurs, conscious that this attitude cannot be sustained because it has no operational support in daily life experience. As a result, instead of asking questions such as "What is life?", or "What is cognition?", or "What is consciousness?" in a way that takes for granted that the answer must arise searching for some support in an external reality in the way we develop our arguments, I began asking questions such as "How do we do what we do as we do whatever we do as human beings?" or "How do we know what we claim that we know?" or "How do we operate as observers making the distinctions that we make in any domain?" in a way that implied that I accepted that the answer that I would accept had to take place in the form of the actual operation of the living systems. And I did so explicitly accepting that all the concepts and notions that I was to use as I answered these questions had arisen derived from the coherences of my living as a living system without introducing any transcendental assumptions in the process. Indeed, to ask these questions as they are presented above entails abandoning *de facto* the implicit metaphysical attitude or *a priori* thinking of a culture that accepts the existence of a transcendental reality as the necessary fundament of all existence, and source of validation of all that we human beings do or can do. Moreover, the very act of asking questions like "How do we do what we do?" in the disposition of answering them as I do, implies accepting that one can answer these questions because they are asked in the domain in which the human beings do what they do as living systems.

A metaphysical attitude that accepts that the essence of being is transcendental entails an attitude that denies the body as the fundament of human knowledge, human understanding, and human consciousness, and gives rise to an epistemological view in which the body is seen as an interference and limitation in the path of true knowledge. At difference from this, a metaphysical attitude that does not arise from the *a priori* acceptance of the existence of a transcendental reality is not concerned with the essences, but instead accepts that all that a human being does arises through his or her body dynamics in the conservation of living in interactions with the medium that makes it possible. From such a metaphysical attitude the body and the body dynamics are recognised by the observer as the fundament of all that the human being does, and the observer asks the questions mentioned above under the general form of "How do we do what we do?" in the full acceptance that our existence as human beings occurs in our relational space in the realisation of our body dynamics. In fact, the implicit or explicit acceptance that we exist as human beings doing whatever we do in the continuous conservation of our human living through our body dynamics is the basic understanding that leads one to abandon the metaphysics of the transcendental reality adopting a new one that takes as starting point for any explanation or rational argument the acknowledgment that we are living systems and do all that we do in the realisation of our living. In this metaphysical view our biology is our condition of possibility. And as a matter of fact it cannot be otherwise since the observer disappears as his or her bodyhood is destroyed.

An example. The metaphysics of the transcendental reality
What is this? – A table. – How do you know that this is a table? – I know because I see it. – And how can you see it? – I can see it because it is there, and I have the ability to see what is there.

This argument stands on an *a priori* explanatory principle that says that something can be distinguished because it is independent of the observer and is independent of the observer because it is real. Moreover, this argumentation stands on the implicit acceptance that there is outside of me an independent reality that is the fundament for all I do, including the reasoning that validates this statement. In this metaphysical attitude a statement is universally valid in relation to what is independent of what the observer does.

A metaphysical attitude arises as a matter of course implicit in the cultural upbringing of a child as an unreflected background of legitimacy that is lived as the ultimate fundament that gives validity to whatever he or she may claim in that culture to be undoubtedly true as a matter of fact or rationally supported. That background is not reflected upon, and if a question arises about its validity such a question is usually answered taking as a fundament for the validity of the answer precisely that which one wants to inquire about. Due to this, if one wants to reflect on the validity of a metaphysical attitude it is necessary that one should release completely the implicit certainty that one has about the nature of the question "What is to know?" and about the manner in which it must be answered. This is what I found myself doing (in my neurophysiological research on visual perception) without being initially aware of what I was doing when I asked in my research on visual perception "What is it to see?"; and I wanted to answer this question looking at the domain of the biological process that constituted seeing in the domain of the operation of the nervous system of the observer in the act of observing as a relational dynamics organism/medium. As I proceeded doing so I soon realised that I had to abandon the notion that the observer existed by itself as an ontologically independent entity, and I realised as well that the question I was asking was about my own operation (How do I do what I do in the domain of seeing?), and that my operations were at the same time what I had to explain and my instruments of explaining them.

I had to explain the observer (myself) and observing (my doing observing) operating as an observer observing, and I had to do so without any ontological assumption about the observing while accepting that the observer arose in its operation as an observer and did not pre-exist its own self-distinction. The task that I began was a circular task and I wanted to explain what occurred in this peculiar circularity (I wanted to explain knowing through knowing) without coming out of it. In doing this I had to explain all that we humans do by doing what we do, not by making any reference to some independent domain of existence. *And all this led me to inquire about living, explaining, language, emotions, and the origin of our humanness.* I was making a metaphysical shift, I moved from the traditional metaphysics that assumes that the world we live pre-exists our living in it, to one in which the world we live exists as it arises with our doing it.

In this metaphysical shift I was abandoning a metaphysical attitude that accepted *a priori* that the observer existed by itself as a transcendental entity that uses other transcendental entities as instruments for explaining and reasoning, and I was adopting one in which the observer arose into existence in the moment of his or her distinction as he or she used as a starting point for all his or her reflection the domain of his or her doings in daily living. In fact I found myself doing this metaphysical shift in the process of explaining the manner of operation of the nervous system in the phenomenon of perception, and before I became aware that in doing so I acted accepting as a matter of course that I, the observer that was doing the explaining, did not pre-exist to his or her own distinction of himself or herself operating in the observing.

An example. The metaphysics of the arising reality

That animal that you see yonder is a horse. – And how do you know that it is a horse? – I know that it is a horse because I recognise in it the characteristics of a horse. – And how do you know that those characteristics that you recognise are the characteristics of a horse? – I know because I have seen them in other horses. – And what is a horse? – An animal that those who know horses call a horse because it has the characteristics of those animals that they call horses. – But that is a circular argument. – No, it is the revelation of the circular operation that constitutes the validation of a distinction in the domain of experiences of an observer as he or she operates as a human being.

In this metaphysical attitude there is no ontological assumption, and the observer is always open to reflect on the fundaments of his or her manner of explaining and on what he or she thinks gives validity to what he or she considers valid. In this metaphysical attitude a statement is universally valid in the domain in which the conditions that give validity to it are satisfied.

This was a fundamental metaphysical change that I did initially without being aware of what I was doing. I was a biologist, a scientist explaining perception and cognition as biological phenomena, and I did not want to lose in the formulation of my explanations the biological processes or phenomena that I was explaining. So I attended at the coherences of my doings and my reflections on my doings in my operation as a human living system. No doubt I was aware that at the same time that I was doing physiology I was aware that to the extent that we all do philosophy when we reflect on the fundaments of whatever we do I was also doing philosophy, but I did not like to say that I was doing so because I did not want to obscure the listening of my colleagues about the scientific nature of my research. Moreover, it was not until Ximena Dávila Yánez, my colleague and cofounder with me of the *Matristic Institute* for teaching Biology of Cognition and Biology of Love in Santiago, said to me that she thought that I had created a new metaphysics, that I became fully aware that in fact I had done so, and that I had to be explicit in acknowledging that I was aware that I was not only doing biology but that I was also doing philosophy. I am grateful to Ximena Dávila Yánez for showing this to me and for the expansion in my understanding that her reflections brought to me.

The separation of science and philosophy is a classificatory artifice that by separating reflection and doing interferes with the understanding of what we human beings do in our living as such, and has obscured our understanding of the different worlds that we bring about in our living, and of what happens to us and in us as we live these different worlds. And this happens because in the separation of science and philosophy we deny ourselves the possibility of fully reflecting on the fundaments of what we do, either because as scientists we think that such reflection is irrelevant because all that matters are facts, or because as philosophers we think that what we need is ultimate truths, and not the pragmatics of material events. The expression 'Natural Philosophy' grasps better what scientists and philosophers want to do when they begin to listen to and look at each other in mutual respect and not mutual devaluation. All that we human beings do occurs in our daily living, and if we do not see and accept that it is so, we cannot appreciate how our biological existence as languaging living systems results in something that could not have arisen through technology without the creative participation of human beings, if for no other reason because technology is a product of our humanness as biological entities. Furthermore, without the metaphysical shift presented in this book, this understanding would not be possible because we would be trapped in the intrinsically unending search for some transcendental reality that we take *a priori* as the ontological ground that is the origin of all that happens to us in our living and our thinking, but which has not and cannot have operational presence.

The doings of daily living are primary in the sense that whether we like it or not they are our starting point for whatever we do or whatever reflections we make. We explain our living with the coherences of our living. However, that we do so does not constitute a circular argument because an explanation does not replace what it explains. Explanations only tell what should happen for that which is explained to arise as a result. So, the explanations of the observer and of observing do not replace the observer or observing, they only show what processes should take place for an observer and its operation in observing to arise, as well as how the observer and the observing would arise if the conditions that give rise to them and to their operation were to take place. Accordingly, it is because the metaphysical shift presented in this book keeps us in the domain of the operational coherences of our living (and all that we do, whatever it may be, we do it in our operation as living systems) that it is possible as a result of this metaphysical shift that we may explain all that we do through the coherences of our living without making any ontological assumption. In a scientific explanation the observer explains his or her experiences with the coherences of his or her experiences mostly unaware of the metaphysical implications of what he or she does. Moreover, scientists frequently argue that their explanations are supported by laws that are a reflection of the coherences of nature as an objective domain of processes that exists fundamentally independent of what he or she does, and they are not aware that the laws of nature are abstractions of the operational coherences of their living.

I was fortunate growing up as a boy who was in some way an unaware natural philosopher interested in understanding the spontaneous dynamic architecture of living beings seduced by their anatomical beauty. And I was also fortunate in that I did so again in an unreflected fundamental attitude of participation in the dynamic architecture of the living because I never saw myself as different from all those marvellous beings that I observed. However, may be that I was not even there different from other children, as I never saw myself different from them in their curiosity, what was again a blessing that permitted me to grow as myself in full respect for being whatever I was becoming.

Finally, I would like to remark in this introduction that although the metaphysical shift that I have done resembles in some aspects to the oriental philosophy, it is fundamentally different. Oriental philosophy stands on the distinction between what is permanent and what is transitory, and invites us to adopt the path of liberation of the transitory to recover the permanent divine essence that we all have. In the oriental philosophy the transitory is an illusion to be overcome. In the metaphysical shift that I have done, in the fundamental attitude of the metaphysics of the arising realities we living systems in general, and we human beings in particular, arise belonging to the domain of the transitory where the transcendental is a notion about something of which we cannot speak because any attempt to do so negates it, and leaves us in the domain of our daily living which is where the transcendental does not exist. Bur this does not matter because all that is good in human life belongs to the domain of what is not permanent, and it is in that domain that love exists as our fundament as human beings, and as our source of well-being.

At the end I wish to express my acknowledgment and thanks to Beatriz Gensch, my wife, for the many conversations that we have had on matters of aesthetics, philosophy, and spiritual life, conversations that have expanded my understanding and enriched my daily living in all dimensions and have brought well-being to me in all that I do. Yet, I want above all to acknowledge that it was through these conversations with her that I became free to talk about love as a scientist.

Humberto R. Maturana
Santiago de Chile, January 2002

I. The cosmos, an explanation of observing

1. Without the observer, there is nothing

EVERYTHING THAT CAN BE SAID IS SAID

POERKSEN: A few pages into your by now legendary paper, *Biology of Cognition*, we come across a treacherously innocent statement that appears to me to be of central importance to your whole work: "Anything said is said by an observer." What does this mean?

MATURANA: What is said can under no circumstances be separated from the person saying it. There is no possibility of validating one's own assertions with regard to an observer-independent reality the existence of which is, in addition, considered as evidently given. Nobody can claim to have privileged access to an external reality or truth.

POERKSEN: But there are innumerable people who would claim that their ideas are true and absolutely valid.

MATURANA: Correct. Nevertheless, all those who think that their assumptions are true in an absolute sense, make a fundamental mistake: they confuse believing and knowing, they claim to have abilities that they, as human beings, simply cannot possess. Of course, it has become accepted in our culture to distinguish between the observer and the observed, or between subject and object, as if there were a difference between the two, as if they were distinct. If this is assumed and accepted, we are immediately confronted by the task of describing the relation between these two supposedly independent entities with greater precision. My contention to the contrary is that this distinction is unhelpful, and I would like to show to what extent all observers are part of their observations.

POERKSEN: What are the consequences of such a position for our ordinary conception of knowledge? Out there, by common sense, exists a world of objects determining what we perceive and describe. What happens to this external reality if we take your key proposition seriously?

MATURANA: The very assumption that this external reality exists independently from us will immediately show itself to be a fundamentally absurd and nonsensical notion: it cannot be validated in any way. There are, of course, philosophers who accept that it is impossible to know that absolute reality but who still insist on its existence. They do not want to give up the certainty of an observer-independent point of reference as part of their background.

POERKSEN: Kant already distinguishes between an absolute reality, the *thing-in-itself*, and a world of appearances. Only the latter, the *phenomenal world*, so he says, is accessible to us.

MATURANA: How can we claim to know of the existence of that absolute reality and at the same time assert its unknowability? This is just an absurd kind of conceptual acrobatics because any talk about that supposedly independent reality is inevitably dependent on the persons talking. When I insist, however, that everything said is said by an observer, I put the spotlight on another key question, which changes the whole traditional system of philosophical discourse about reality, truth and the essence of being. This question is no longer concerned with the investigation of an external reality that is taken to be given and to exist independently from us. It is the observer whose operations I want to understand by operating as an observer; it is language I want to explain by living in language; it is speaking I want to describe more precisely by speaking. In brief: There is no way of approaching what we want to explain from outside ourselves.

POERKSEN: The immediate consequence of what you are saying is that the strict division between an external world and a knowing subject collapses: the situation turns circular.

MATURANA: This is the decisive point. The subject and, at the same time, inevitably, the instrument of my inquiry is the observer. We are indeed entwined in a circular situation that replaces the traditional separation of the observer and the observed. I am not interested in the question as to whether an observer-independent reality exists and whether I or somebody else may know it. I use the observer as the starting-point of my thinking, avoiding any ontological commitment, simply out of curiosity and interest in the questions involved. There is no higher reason, no ontological foundation, no universally valid justification. The observer observes, sees something, affirms or denies its existence, and does whatever he does. What exists independently from him is necessarily a matter of belief and not of certain knowledge because to see something always requires someone who sees it.

POERKSEN: A closer consideration of your key aphorism makes me feel slightly uneasy. Such a postulate seems so categorical and irrefutable. Of course, and that is immediately obvious, everything said is said by an observer. There is no way around this insight; it seems ineluctable. So: Under what circumstances could that statement be refuted?

MATURANA: Only God could do it. God could talk about everything without observing it because S/He *is* everything. We do not possess God's abilities as we are forced to operate as human beings. Nothing can be said without some person saying it.

POERKSEN: This would mean – following Protagoras –: The observer is the measure of all things.

MATURANA: My claim is even stronger: The observer is the source of everything. Without the observer, there is nothing. The observer is the foundation of all knowledge, of any assumption involving the human self, the world and the cosmos. The disappearance of the observer would mean the end and the disappearance of the world we know; there would be nobody left to perceive, to speak, to describe, and to explain.

In the beginning was the difference

Poerksen: How can you be so sure that nothing exists without the observer? A postulate like that could easily be interpreted and judged as the presentation of a new kind of truth. Then you would contradict yourself.

Maturana: There is no question of a new kind of truth. By focussing on the observer and the operation of observing I want to introduce a topic of research and, at the same time, characterise a way of dealing with it. We must be quite clear about the fact that the very idea of something given and existing, and the very reference to some reality or some sort of truth, unavoidably involves language. Whatever we are able to say about that truth or reality is dependent on the availability of language. What is supposedly independent from us becomes describable only when language is available, emerges only through an act of distinction by means of language. Even for the process of meditation, when we believe that we are moving in a state of pure consciousness, we are forced to admit that the reflection of that state cannot be achieved without language.

Poerksen: Is your claim then that we cannot escape from language, that we can never get out of our linguistic universe?

Maturana: Language is not a prison; it is a form of existence, a way and a manner of living together. The simple expression 'we cannot *escape* from language' makes us believe that there *exists* some other place, a place *beyond* language, although it may never be reached. I refuse to make such an assumption. Living in language implies that it is meaningless to think about a world that might exist beyond language. Just consider the comparable question: If everything is part of the universe, can we still somehow get out of the universe? The answer must be: The universe is wherever I go. We are inseparably moving together.

Poerksen: But is your key concept of the *observer* not an unfortunate choice? In ordinary language, it signals separation: We observe, keep a distance, and indirectly insist on neutrality. Would it not be advisable to replace observers by *participants* who are inseparably tied to their worlds?

MATURANA: I am not at all unhappy with the concept of the observer because the way we speak about things in our daily experience naturally implies that the things we perceive and handle exist independently from us. We even speak about ourselves as if we were separated from ourselves, as if we were able to observe ourselves from an external point of view. This is to say: Observers are human beings who distinguish something – including themselves – as if it could be separated from them. And this experience must then be explained.

POERKSEN: If I understand correctly, it is one of your goals to find out why we experience something as separate from us, in the first place.

MATURANA: Precisely. Therefore, the suggestion to speak of *participants* does not appeal to me. The notion of participation is misleading because it contains an explanation and a ready-made answer; the only admissible question left would then concern the specific manner in which the assumed participation is realised. The table and the chairs in this room, my jacket, the scarf I am wearing – all these things undoubtedly appear to exist independently from me: we think we are outside the given situation and separated from it. This means that observing is an experience which also has to do with the apparently independent existence of things. The problem, therefore, is: How do I know that these things are there, at all? What sort of claim is it to say that the world unfolding before my eyes exists independently from me?

POERKSEN: So your starting point is the experience of separation in order to establish and justify the insight that we are inevitably involved in the construction of our realities and therefore tied to them.

MATURANA: At the outset, there is the experience of separation, which, finally, turns into the insight of connectedness. Of course, I am not part of the object I am describing; I am not part of the glass here on the table when I am pointing at it. However, the distinction

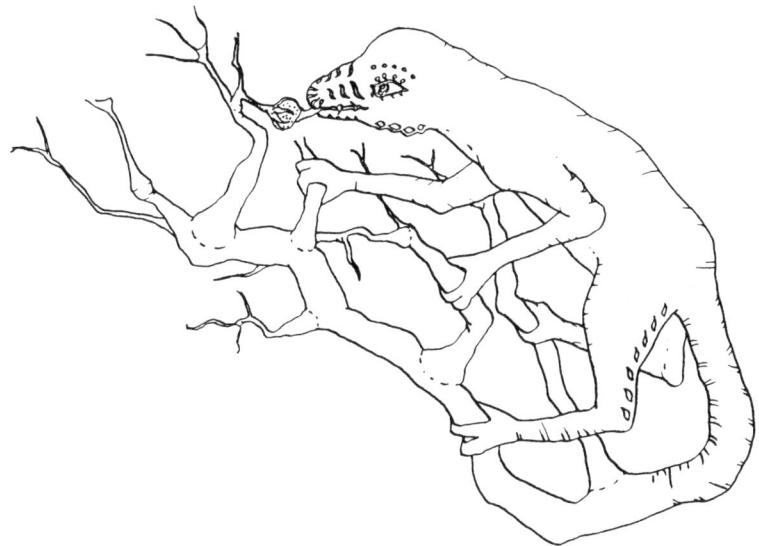

Fig. 1: The tree of knowledge: No experience of living beings is independent from them (drawing by Marcelo M. Maturana).

of the glass has to do with me; I am the one who describes it, I am the one who uses the distinction. Or the other way round: If nobody makes this distinction then the material or conceptual entity that is specified and demarcated from an environment in this way does not exist.

POERKSEN: The first distinction we make is, therefore, something like the Big Bang of knowing, the origin of the construction of a reality. There must be a distinction in order to be able to see anything at all.

MATURANA: Exactly. Only what is distinguished exists. Although it is distinct from ourselves, we are nevertheless tied to it through the operation of distinction. Whenever I distinguish something, the entity that is distinguished emerges together with some background in which the distinction makes sense; it brings forth the domain in which it exists.

POERKSEN: Could you be a little more specific and give an example?

MATURANA: Just imagine the following situation. One evening you visit friends to celebrate a party. Suddenly, in the middle of a conversation with a couple of people, someone touches you on the shoulder. You turn round and recognise a friend whom you have not seen for many years. Your friend seems to emerge out of nowhere. "Oh," you say, "What are you doing here?" You ask him where he comes from, who invited him, what his life is like, etc. In fact, you create a history, a domain of connections, a background giving meaning to his appearance. In this way, his sudden emergence out of nowhere loses its frightening strangeness.

THE EXPLANATION OF EXPERIENCE

POERKSEN: If the perceptions of observers are dependent on their distinctions, then the world constructed by these observers might conceivably not exist at all. Furthermore, all the other human beings around them might be considered mere figments of their imagination, chimeras of isolated minds. This is the conception of epistemological solipsism. Would you agree with the solipsists?

MATURANA: No, not at all. The simple reason is that I do not experience myself as alone and isolated. On the contrary, I experience myself sitting in this house together with you and the two of us talking to each other. It is this sort of experience, whether my own or someone else's, that forms the point of departure of all my further reflections and explanations. Therefore, I cannot take a solipsistic position; such a categorisation of my thinking would be totally misleading.

POERKSEN: Obviously, you are not alone now. We are doing an interview. Is it the experience of the interview that saves you from slipping into solipsism?

MATURANA: Exactly. However, we must now raise the question of how we can explain the experience of being together with someone else if we cannot distinguish anything that is independent from us. The answer for me is that language is a way and manner of living together. – Who is living together? My reply: Human beings. The next question is: What are human beings? I say: Human beings are those special entities that are distinguished in the process of human beings living together. Again, we find ourselves in a circular situation. For me, a human being is not an ontic or ontological entity, an entity existing *a priori*.

POERKSEN: Nevertheless, if you do not conceive of your fellow human being as something given, then the interviewer who is sitting at your table right here and now might just as well be an illusion, a mere phantom in your mind. And you would be a solipsist after all.

MATURANA: This is not the necessary consequence. Of course, I could come to the conclusion that you are an illusion and that I am only imagining and presuming your presence, but this would not necessarily make me a solipsist. Although you might be an illusion I would not inevitably be a solipsist because I spend my ordinary life together with my wife: her existence does not have the status of an illusion for me, at all.

POERKSEN: Is it not conceivable, however, that your wife and the rest of the world do not really exist?

MATURANA: If we believed that we are all mere illusions, this would be of no consequence whatever. Our conversation would then have no basis. To be able to classify an experience as an illusion requires that this experience can be related to something else that is not experienced as illusory at the same time. I can only repeat: My starting point is my experience, and that means, all that I experience and distinguish as perceivable events at a certain point in time. I am not concerned with the existence or the properties of an external reality, nor with the defence of solipsism or any other kind of epistemology. I want to understand and explain the operations that generate and form our experiences. In the very act of explaining these operations it becomes evident that we emerge ourselves as the objects and entities we describe.

POERKSEN: You are not a solipsist, and certainly not a realist. In Germany, at least, most people take you to be a constructivist who represents a position in the middle between two epistemological extremes. The classic type of constructivism, however, assumes that there is an external, even an absolute reality but that we are in no way capable of knowing the intrinsic and true form of that reality. It is only when our constructions fail and collapse that we can realise that they were wrong, that they did not fit reality.

MATURANA: I do not share this view, either. How can one show that the clash between my constructions and reality, which proved the constructions to be wrong, has actually taken place? What validity has such an assumption; how could it be confirmed? For me, the collapse of a hypothesis is nothing but an event that disappoints our expectations. In brief: I do not consider myself a representative of constructivism, even if I am called a constructivist over and over again.

POERKSEN: What would you call yourself then? What sort of label would best characterise your position?

MATURANA: I hesitate with my answer because such a label may affect the perception and appreciation of what I am saying in a negative way; if you are labelled you are not seen. However, whenever I am asked for a suitable label, I sometimes call myself – in earnest, but playfully – a "super-realist" who believes in the existence of innumerable equally valid realities. Moreover, all these different realities are not relative realities because asserting their relativity would entail the assumption of an absolute reality as the reference point against which their relativity would be measured.

THE AGE OF SELF-OBSERVATION

POERKSEN: My thesis is this: We live in the age of self-observation. It has become fashionable to reflect constantly one's emotions and thoughts, one's feelings and beliefs, and to ponder their variability. Could this craving for life-long therapy be a reason for the enormous popularity of your observer theory?

MATURANA: Possibly, although it would be a total misunderstanding to believe that I am proposing or in any way recommending constant self-observation simply because I am speaking about the operation of observing. In that case, I would have become well known due to a misleading interpretation of my work, which cannot, of course, be ruled out. In my view, however, the real wisdom of a person does not consist in perpetual self-examination but in the capability of reflection, in the willingness to give up those beliefs, which stand in the way of an accurate perception of specific circumstances. The wise do not constantly observe themselves, they do not cling to things, and they do not allow themselves to be guided by ultimate truths that prescribe how they themselves or other people have to act.

POERKSEN: A terminological point: What, in fact, is an observer? How would you define the concept?

MATURANA: Observing is for me a human operation that requires language together with the awareness that one is engaged in observing something. A cat just eyeing a bird does not to me appear to be an observer. It simply watches the bird and, as far as we know, is incapable of commenting on its action or asking itself critically whether it is acting in the right and proper manner; this cat may, from our point of view, behave adequately or inadequately, but it does not reflect its own behaviour. Only human beings can do this.

POERKSEN: Observing is self-reflection.

MATURANA: Exactly. Observers act in self-awareness, when they use a distinction in order to distinguish something; they are mindful when they see and perceive something. Somebody who is simply looking out the window I would not consider to be an observer. The consequence, therefore, is that most of the time in our lives we do not operate as observers; we just carry on without bothering to examine what we are doing.

POERKSEN: In your books you speak about a *standard observer* and a *super-observer*. Do you mean to suggest that there are different degrees of understanding?

MATURANA: No, this distinction must be interpreted differently. When I formulated it, I was possibly grappling with the description of operations of observing that are identical but nevertheless differ in a certain way. Whenever we observe something, we are all standard observers. However, as soon as we ask ourselves what we are actually doing in that moment, we find ourselves in a different situation and position, although, of course, remaining standard observers; we might say that we become meta-observers. Such meta- or super-observers treat themselves as objects and observe – operating as observers – their own observations.

POERKSEN: "Objectivity is a subject's delusion," we read in an address by the biophysicist Heinz von Foerster to the *American Society for Cybernetics*, "that observing can be done without him. Invoking objectivity is abrogating responsibility – hence its popularity." You worked with Heinz von Foerster in the late sixties. How do you interpret these statements?

MATURANA: They deal with the belief that observations can be separated from what is observed, that the person of the observer is marginal and can easily be replaced because observing is a simple process of merely registering what is happening; the observer's own actions are lost sight of. What is used to confirm some statement comes from outside, it is related to reality or truth. The foundation of any judgment appears to be external to the person of the observer. The usual conclusion is, therefore, that no one can be held responsible for those judgments, as they seem to have nothing to do with personal predilections and interests.

POERKSEN: It seems to me that your reflections point in the opposite direction: We become aware of our responsibility for our perceptions and our claims.

MATURANA: Correct. Becoming aware that one is doing the observing, and then being aware of being aware that it is oneself who makes the distinctions, one attains a new domain of experience. Only becoming aware of one's awareness and understanding one's understanding can give rise to the feeling of responsibility for what one is doing, for what one is creating through one's own operations of distinction. This kind of insight has something inevitable: Once this has been understood, one cannot pretend any longer to be unaware of one's own understanding if one is actually aware of it and is also aware of this awareness. Moreover, the concept of the observer is a challenge to study the operation of observing and to face up to the circularity of the understanding of understanding. It is, after all, an observer who observes the observing, it is a brain that wants to explain the brain. Many people think that such reflexive problems are unacceptable and unsolvable. My proposal, however, is to accept this circular situation fully right from the start and to make oneself the instrument by means of which the question of one's personal experience and one's own actions is to be answered through one's very own activities. The point is to observe the operations, which give rise to the experiences that are to be explained.

2. Varieties of objectivity

LIFE IN THE MULTIVERSE

POERKSEN: Your plea for circular thinking somehow seems deeply disturbing, even threatening. The world dissolves; beginning and end become arbitrary fixations no longer offering a safe grip; all firm ground is pulled from under our feet. One would like to rush to the door and out of the room, but one has become uncertain that that door is still there. You reported somewhere that, having started to think in this way, you were quite scared for some time that you might go mad. Why did that alarm finally fade away?

MATURANA: There came a moment at which I realised that circular thinking did not endanger the soundness of my mind but that it expanded my understanding. The decision, in particular, to proceed from my own experience and not from an external reality can have a profoundly liberating and comforting effect. The experiences we have are no longer doubted, no longer denigrated as irreal and illusory; they are no longer a problem, they no longer produce emotional conflicts; they are simply accepted for what they are. – Suppose I claim to have heard the voice of Jesus speaking to me last night. What do you think would happen if I told other people of such an experience? Somebody might explain to me that I suffered from hallucinations because Jesus was dead and could therefore not possibly speak to me. Someone else might think me very vain and suspect that I wanted to present myself as an elect person: it is, after all, Jesus who was been speaking to me. A third person might say that during that night the devil had led me into temptation. All these considerations have one thing in common: they reject the explanation with which I am trying to make sense of my experience but they do not negate the experience itself; they do not call into question that I heard a voice.

POERKSEN: In what way does this example contribute to answering my question concerning your fear of madness? I assume that your decision to start out from your own experience allayed your fears, calmed your mind, and set you at ease. One accepts what one experiences. Therefore, the fear of madness might be a sort of clandestine attempt to defend oneself against one's own experiences.

MATURANA: Exactly. To call something *mad* means to explain one's perceptions and experiences in such a way as to devalue oneself. It is not my intention to reject or devalue experiences. Experiences are not the problem. What I want to explain is the operations through which experiences arise.

POERKSEN: Do you believe that such a view, which so forcefully argues in favour of the legitimacy of any kind of experience, offers ethical advantages?

MATURANA: Yes, I do. We must not forget that the notion of a reality existing independently from us corresponds with the belief that it is possible to achieve authoritative, universally valid statements. These may be used to discredit certain kinds of experience. It is the reference to this reality that is held to make a statement objective and universally valid; in a culture based on power, domination and control, it provides the justification for forcing other people to subject themselves to one's own view of things. However, as soon as one has realised that there is no single privileged access to reality, and that perception and illusion are indistinguishable in the actual process of an experience, then the question arises what criteria can be used by a human being to claim that something is the case. The very possibility of posing this question opens up a space of common reflection, a sphere of cooperation. The other person becomes a legitimate other with whom I am able to talk. Friendship, mutual respect, and cooperation emerge. It is no longer possible to demand submission; the universe changes into a multiverse within which numerous realities are valid by reference to different criteria of validity. The only thing one may now do is to invite the other

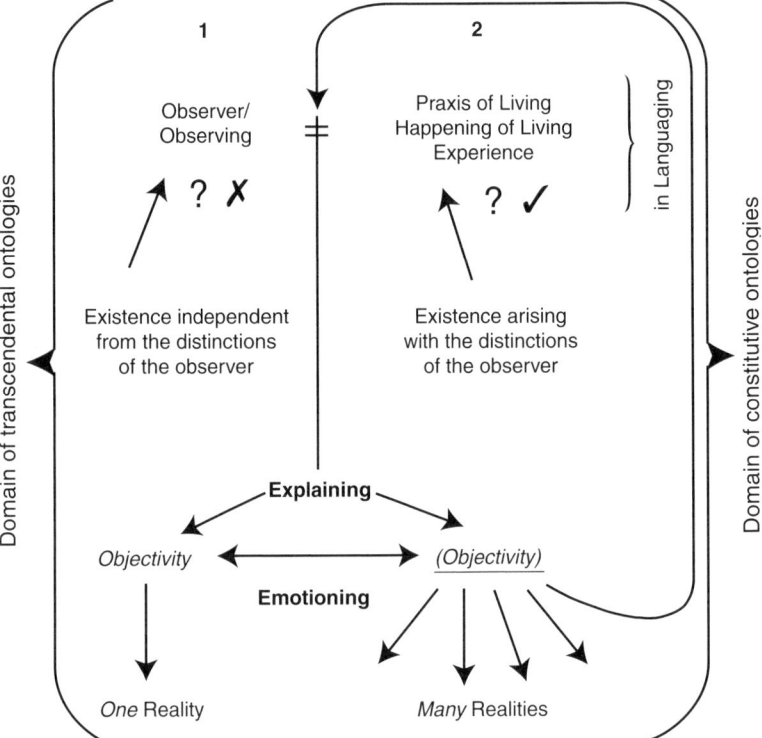

Fig. 2: Diagram of the ontology of the observer. The diagram shows what happens in the case in which one accepts the question about "how we do what we do as observers observing" (explanatory path 2), and what happens if we do not accept that question (explanatory path 1). If one knows how to read this diagram, the understanding of how the observer arises as a biological entity unfolds, showing that the observer is a manner of relational operation that takes place in the realisation of the living of the human being as a languaging living system. This diagram is in the domain of cognition like the formula $E=mc^2$ in the domain of physics.

person to think about what one believes and holds to be valid oneself.

POERKSEN: This means that we have two fundamentally different positions. One claims that all knowledge is observer-dependent, the other, that an observer-independent reality can be perceived. Both positions have their different consequences and lead to specific approaches to the environment and to other people.

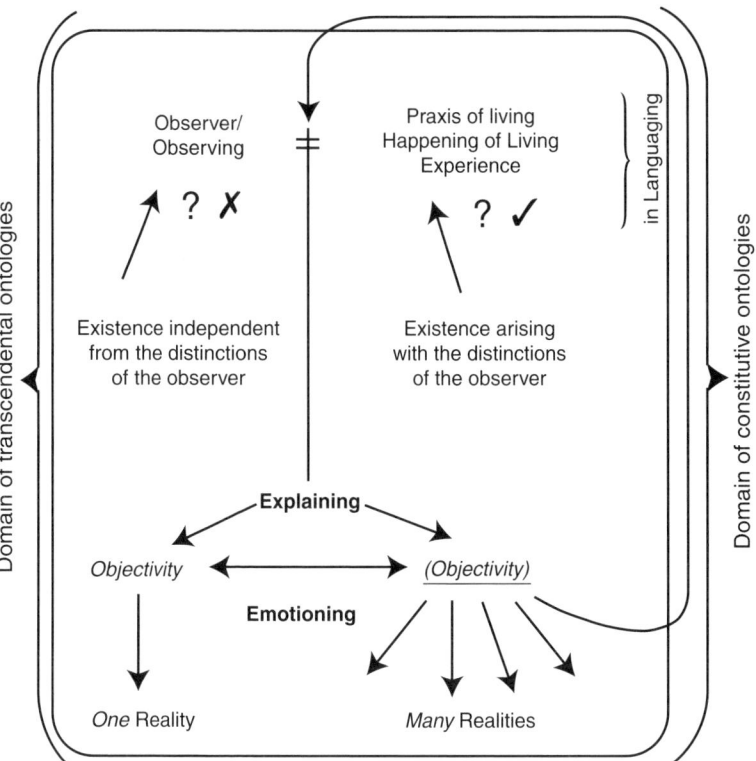

Fig. 3: Diagram of the understanding of the biological matrix of human existence. This diagram shows the understanding of the relational dynamics of the constitution and conservation of humanness in the historical interplay of the biology of cognition and the biology of love that Ximena Dávila Yánez and Humberto R. Maturana have recently developed and which they call the Biological Matrix of Human Existence. Essentially it reveals the different paths of awareness of our relational human existence that we can live according to the flow of our emotioning and also reveals our basic plastic being open to change through reflection.

MATURANA: There are two distinct attitudes, two paths of thinking and explaining. The first path I call *objectivity without parentheses*. It takes for granted the observer-independent existence of objects that – it is claimed – can be known; it believes in the possibility of an external validation of statements. Such a validation would lend authority and unconditional legitimacy to what is claimed and would, therefore, aim at subjection. It entails the negation of all those who are not prepared to agree with the "objective" facts. One does not want to

41

listen to them or try to understand them. The fundamental emotion reigning here is powered by the authority of universally valid knowledge. One lives in the domain of mutually exclusive transcendental ontologies: each ontology supposedly grasps objective reality; what exists seems independent from one's personality and one's actions. The other attitude I call *objectivity in parentheses;* its emotional basis is the enjoyment of the company of other human beings. The question of the observer is accepted fully, and every attempt is made to answer it. The distinction between objects and the experience of existence is, according to this path, not denied but the reference to objects is not the basis of explanations, it is the coherence of experiences with other experiences that constitutes the foundation of all explanation. In this view, the observer becomes the origin of all realities; all realities are created through the observer's operations of distinction. We have entered the domain of constitutive ontologies: all Being is constituted through the Doing of observers. If we follow this path of explanation, we become aware that we can in no way claim to be in possession of *the truth* but that there are numerous possible realities. Each of them is fully legitimate and valid although, of course, not equally desirable. If we follow this path of explanation, we cannot demand the subjection of our fellow human beings but will listen to them, seek cooperation and communication, and will try to find out under what circumstances we would consider to be valid what they are saying. Consequently, some claim will be *true* if it satisfies the criteria of validation of the relevant domain of reality.

A MULTITUDE OF WORLDS

POERKSEN: Your conceptual differentiation seems to me to be a little too complicated. Why not simply distinguish between *objectivity* and *subjectivity* in order to separate these two positions?

MATURANA: 'Subjectivity' is one of the expressions that we use to devalue a statement from the path of *objectivity without parentheses*. 'Purely subjective' we call a statement, which does not correspond with reality. By speaking of *objectivity in parentheses,* I want to keep everybody aware of the fact that it is impossible to establish an observer-independent point of reference. At the same time, I want to conceptualise our experience that there seem to be objects independent from us. The parentheses are meant to signal a certain state of awareness. My question is: How can it be that we experience objects as distinct from us although we know that everything said is said by us and cannot be separated from us?

POERKSEN: Hearing you speak like that about conceptual distinctions, I am beginning to get a clearer idea of a principle guiding your use of language: your terminology with all its neologisms is firmly based on the experiences of human observers but at the same time suggests a different view of those experiences.

MATURANA: That is the idea, precisely. From time to time, I have been criticised for still talking about ontology and existence; I was reproached for not replacing ontological considerations by an ontogenetic perspective because people thought it was essential to focus on processes of becoming. Such a demand appeals to me, of course, but the concomitant rejection of reality and of the objects that unquestionably manifest themselves in the actions of observers, negates the ordinary experiences we human beings make every day. It cannot, therefore, serve as a reliable basis for my argumentation.

POERKSEN: If we remain aware that everything said inevitably refers back to an observer, our universally valid reality breaks up into innumerable different realities. More than six billion people live on this earth; would you say that there are more than six billion realities?

MATURANA: It is a theoretical possibility but factually quite improbable. If we assume that about five billion out of the six billion people follow the path of objectivity without parentheses, they live more or less in the same domain of reality: some of them believe in Allah, others in Jehovah or Jesus, still others are agnostics. Some of them consider consciousness as the unconditionally valid reality, others matter, or energy, still others are in favour of ideas and notions as the ultimate points of reference for their claims. However, they are all united by one fundamental certainty: *they do not believe that they are believing but they believe that they know because they do not know that they are believing.*

POERKSEN: What about the remaining billion people? How would you characterise their attitude?

MATURANA: They might – possibly – follow the path of objectivity in parentheses and, therefore, be capable of reflection: They would respect differences, would not claim to be the sole possessors of truth, and would enjoy the company of others. In the process of living together, they would produce different cultures. Consequently, the number of possible realities may seem potentially infinite but their diversity is constrained by communal living, by cultures and histories created together, by shared interests and predilections. Every human being is certainly different but not entirely different.

POERKSEN: Can I live in the awareness that there is a potentially infinite number of different realities? I would suspect that any attempt to exhaust the infinite multitude of possible worlds must inevitably lead to breakdown and the complete loss of orientation.

MATURANA: Of course, we need to reduce complexity; we need to narrow down our focus and rely on specific expectations to be capable of acting at all. However, the problem is not sticking to certain expectations, reducing complexity, or categorising a multitude of phenomena under one, or even only one, concept. The central problem for me is whether one is prepared to give up one's certainties when something unexpected develops. Disappointing experiences need not necessarily lead to deep frustration and anger but may, quite undramatically, open up new perspectives. One realises that one's expectations are not fulfilled and, without great excitement, decides on a new orientation.

POERKSEN: How does one learn to move in the world in this way? How does one acquire the awareness that – although one has already chosen a particular variant from the multitude of possible ways of life – everything could be quite different?

MATURANA: Certain events in one's life bring about insights of this kind. For example, it happens often enough that one has a certain belief and then encounters a person that one would have to reject according to that belief. One should not, in fact, like the person, but one simply does, and so one realises that one's beliefs and the sympathetic view of the person do not fit and cannot both be upheld any longer. If one's beliefs are given priority, then the person in question will disappear as a likeable human being from one's field of vision. However, if one chooses to give in to the attraction, then one begins to reflect one's judgments and their effects and bids them farewell. In such ways, we learn how crippling the effects of beliefs and certainties of all kinds may be; they impose a kind of perception that we ourselves on reflection find inadequate.

POERKSEN: Certainties are, therefore, essentially dangerous as far as their consequences are concerned: they render alternative ways of feeling, thinking and acting invisible.

MATURANA: If they do not just surface as transitory certainties of the moment, they are something extremely powerful. They make us blind and make all further reflection seem a waste of time: We believe we already know in advance the only possible result of any renewed reflection effort. What, in fact, do we really mean when we say we are absolutely sure of something? We say: There is no point in entertaining doubts; our beliefs are so overwhelming that it must appear completely absurd to think about the conditions of their origins. Immediate action seems required. And should we, in addition, want to free others of their supposed ignorance and their false perception of the world, we would become a real danger: the authority of reality will then serve as an instrument to justify exploitation and subjugation, wars and crusades.

POERKSEN: Would you say that certainties and a belief in absolute truth necessarily lead to the suppression of other ways of thinking?

MATURANA: Sometimes I think that we live in a culture where the belief that one is in the possession of truth is understood as an invitation to imperialism. Why should we, who definitely know what is correct, allow the others to go on living in ignorance? Would it not be better, appropriate and, in fact, indispensable, people ask in this culture, to put that allegedly false view of the world right and replace it by the true and correct one? At some stage, everything unfamiliar and extraordinary will, consequently, appear as an unacceptable and insupportable threat and its correction and elimination will be deemed appropriate. Everybody knows what the facts are; everybody knows the right answers, the right ways of living, the true God. The possible consequence of such an attitude is that people feel justified to use violence because they claim to have privileged access to *the truth* or to fight for a great ideal. This attitude, so they believe, justifies their behaviour and sets them apart from common criminals.

POERKSEN: Who is the target of this criticism of an idea of truth turned totalitarian? Where do you see such forms of conflict?

MATURANA: They are ubiquitous although they need not always end in physical terror. In political and polemical debates, which are often similar to fights or even wars, we reject other people and their views. We attack them without listening, we, in fact, refuse to listen because we are sure that they hold views that are false. Political terrorism rests on the idea that certain people are wrong and must, therefore, be killed.

TOLERANCE AND RESPECT

POERKSEN: Is there not a less dangerous and less fanatical way of handling the view that one has discovered the reality of the world as it is?

MATURANA: It all depends on the emotions of the people related to each other. If they respect each other, then the fact that they hold different views may offer the opportunity of a fruitful conversation and a productive exchange. If people, however, demand subservience, then the differing views will provide motives for negation.

POERKSEN: If we train ourselves, as you suggest, to recognise the abundance of life forms and to feel at home in a multiverse, we still face the necessity of choice: We cannot accept everything, we have to choose, to decide on some kind of existence, and to limit the infinity of possibilities. This is easy for ordinary realists: They simply insist that it is the objective necessities that dictate what they decide. You would no doubt reject such an argument. Therefore: What is your criterion for taking the necessary decisions?

MATURANA: We do what is good for us, what sustains and improves our well-being. Take the man, for example, who wants to train as a cook. Why a cook? "Well," he says, "cooks are needed – so I shall have work and a comfortable way of supporting myself; and I love cooking!" If you listen carefully, you will realise that the reasons he gives all have to do with sustaining and improving his well-being. This is not a plea for hedonism, not at all; it is simply my suggestion to listen carefully to people telling you about their life decisions. The cook-to-be will certainly add that one can make a lot of money in his trade; but this only means that for him well-being seems to depend on income.

POERKSEN: This criterion of well-being seems to suggest that we should simply accept any imaginable decision people take with regard to their course of life. Are you advocating absolute tolerance?

MATURANA: For me, the plea for tolerance has an extremely unpleasant flavour; it is an expression of an inclination towards the path of *objectivity without parentheses*. People who demand tolerance are actually only proposing to delay and suspend for a little while the rejection and debasement of other people, which they have already decided to be justified. People who merely tolerate other people will leave them alone for some time but always have the knife ready hidden behind their backs. They do not listen to the other people, do not really give them their attention; their own ideas and beliefs remain in the foreground. The others are in the wrong but their destruction is postponed for a little while: that is tolerance. Following the path of *objectivity in parentheses*, however, we meet other people's worldviews with respect, we are prepared to listen to them, to acknowledge their realities and accept them as fundamentally legitimate.

POERKSEN: When do realities definitely become unacceptable – even for those who believe in *objectivity in parentheses?* Under what conditions must fundamental respect end?

MATURANA: Respect never ends. If we realise, of course, that certain people are creating a world that we consider dangerous and highly unpleasant, then we will certainly act and stand up to them because we do not want to live in their world. I think that this kind of justification of one's actions is crucial: We no longer appeal to a transcendental reality or truth in order to provide grounds for our actions but we act in full awareness of our own responsibility. We do not like nor want the world we see and, therefore, we become active and reject people in a responsible way, or bring about a separation in mutual respect.

POERKSEN: Could you be a little more specific with regard to the somewhat unusual distinction between tolerance and respect that you are proposing here? These two concepts are normally thought to be identical and used as synonyms.

MATURANA: Right, but that is a colossal mistake. Perhaps an example can enlighten us here: Churchill had great respect for Hitler – and therefore understood Hitler's real intentions, which made him oppose National Socialism. Chamberlain, however, showed enormous tolerance towards Hitler and, therefore, proved incapable of assessing the man properly, which made him negotiate completely foolish contracts with him.

POERKSEN: Consequently, such an attitude of respect might very well make us decide one day – in full awareness of our responsibility – to make use of a gun?

MATURANA: Certainly. People might read *Mein Kampf* and realise immediately that, in this book, Hitler quite openly reveals his intentions and goals. They will then have to decide whether they really want to support the world described there and the programme laid before them. It is the respect for other persons' realities which enables us to assess them properly and to take mindful action: We listen to them in order to decide. – People who tolerate their enemies do not see them because their own beliefs obscure their perception; respecting the enemies, however, makes it possible to understand them – and to stand up to them if necessary.

POERKSEN: For me, the question is now how we might promote and practise this very fundamental kind of respect in a manner that does not in any way involve domination. If you want to remain consistent, you surely cannot force other people to agree to your thoughts. How are we to proceed, then, if dominance and manipulation are inadmissible? How do you convince people?

MATURANA: I never attempt to convince anyone. Some people become annoyed when they are confronted with my considerations. That is perfectly okay. I would never try to correct their views and then force my own ideas upon them. Other people, however, are touched by what I have published during the last few decades because they find it affects their own life. They do not merely read what I have written but come to my lectures, which are invitations to follow my reflections. – The only thing left for me to do is to converse with those people who seek and wish to converse with me. I give lectures if people want to listen to me; I write articles and books and work with my students. And one day perhaps a young man might come to Chile from Germany to visit me and ask for more precise details.

POERKSEN: You say that you invite people to listen to you. An invitation, however, has a great disadvantage if quick action is required: it may by definition be rightfully declined. However, people who proclaim laws and formulate imperatives have the enormous benefit of speed; given the necessary power, they may gain quick control over people and rapidly align them with regard to their own goals and purposes. Perhaps invitations may sometimes simply take too long.

MATURANA: What would the alternative be? Should we put people in prison and chains in order to demonstrate to them the wonderful advantages of freedom? Can we force people to reject violence? Such an approach will never work. My view is that even so-called ethical laws and imperatives destroy the possibility of reflection: They remove the foundations of personally responsible action and demand submission; they are, at closer inspection, just another expression for tyranny. You can show people what will happen if a certain worldview or way of life is chosen; you can present them with the consequences implicit in their beliefs and actions but that is entirely different from forcing them to do something and pressuring them, more or less violently, into accepting a particular view of things.

Aesthetic Seduction

POERKSEN: You too plead for a new kind of thinking, for a more respectful form of living together, and you try, at the same time, to show unconditional respect to those people who do not want this change at all.

MATURANA: Of decisive importance is a change in awareness, which cannot, however, be brought about by force in any way; it must emerge through the insight of every human individual. There is no point in denying that I would definitely want a different kind of world even though the simple thought of wishing to change not only oneself but also other people inevitably confronts one with the temptation of tyranny. Of course, I long for a world made up of democratic communities with cooperative individuals respecting one another. I would like to contribute to such a form of living together, which can only come about without pressure and violence, and all I can do is to act as a democratically inspired individual in order to support and keep democracy alive. This means: the journey is the destination; the means I have available are an immediate expression of the goal I wish to reach. Nobody can be forced to accept democracy, nobody.

POERKSEN: You are in the fortunate position that people in the academies and universities of the world are willing to listen to you. What would happen if people no longer wanted to listen to you? What would you do then?

MATURANA: What would happen then? That is all perfectly legitimate. In some of my lectures I mention that I have added three further rights to the United Nations catalogue of human rights: the right to make mistakes; the right to change one's view; and the right to leave the room at any moment. If people are allowed to make mistakes, they can correct them. People who are entitled to change their views can reflect. If people have the right to get up and leave at any moment, they will stay only if they wish to.

POERKSEN: In the last passages of your paper *Biology of Cognition* you outline the concept of *aesthetic seduction*. What does this mean? How can one use beauty and aesthetics to persuade and convince in an appealing manner?

MATURANA: The idea of aesthetic seduction is based on the insight that people enjoy beauty. We call something *beautiful* when the circumstances we find ourselves in make us feel good. Judging something as *ugly* and *unpleasant*, on the other hand, indicates displeasure because we are aware of the difference to our views of what is agreeable and pleasant. The aesthetic is harmony and pleasure, the enjoyment of what is given to us. An attractive view transforms us. A beautiful picture makes us look at it again and again, enjoy its colour scheme, photograph it, perhaps even buy it. The relationship with a picture may transform the life of people because the picture has become a source of aesthetic experience.

POERKSEN: It would interest me to know what this idea of aesthetic seduction means to you when you write, give lectures or interviews. Although this sounds like probing for rhetorical tricks and manipulation, I would like to know what you are, in fact, doing when you try to seduce people.

MATURANA: I certainly never intend to seduce or persuade people in a manipulative way. Beauty would vanish if I tried to seduce in this way. Any attempt to persuade applies pressure and destroys the possibility of listening. *Pressure creates resentment.* Wanting to manipulate people stimulates resistance. Manipulation means exploiting our relation with other people in such a way as to give them the impression that whatever happens is beneficial and advantageous for them. But the resulting actions of the manipulated person are, in fact, useful for the manipulator. Manipulation, therefore, really means cheating people.

POERKSEN: What should we do then?

MATURANA: The only thing left to me in the way of aesthetic seduction is just to be what I am, wholly and entirely, and to admit no discrepancy whatsoever between what I am saying and what I am doing. Of course, this does not at all exclude some jumping about and playacting during a lecture. But not in order to persuade or to seduce but in order to generate the experiences that produce and make manifest what I am talking about. The persons becoming acquainted with me in this way can then decide for themselves whether they want to accept what they see before them. Only when there is no discrepancy between what is said and what is done, when there is no pretence and no pressure, aesthetic seduction may unfold. In such a situation, the people listening and debating will feel accepted to such an extent as to be able to present themselves in an uninhibited and pleasurable manner. They are not attacked, they are not forced to do things, and they can show themselves as they are, because someone else is presenting himself naked and unprotected. Such behaviour is always seductive in a respectful way because all questions and fears suddenly become legitimate and completely new possibilities of encountering one another emerge.

3. The biology of cognition

THE EXPERIENCE OF TRUTH

POERKSEN: You say that all knowledge is necessarily observer-dependent, that absolute reality assertions lead to terror, and that any form of coercion must be rejected. My impression is that all the ideas we have been discussing so far involve ethical assumptions in a very wide sense. We have been talking about conclusions and consequences relating to the claim that objective knowledge of what is real must be impossible. My question is now whether your ethical demands can be justified epistemologically. Is there conclusive evidence for the presumption that truth must remain unknown forever? Is there proof?

MATURANA: Answering your question requires the clarification of what we want to accept as proof. What does it really mean to say that something is true or false? Is a hypothesis proved because it fits into what I am thinking? Am I perhaps prepared to listen and to trust the method of proof simply because of this correspondence between the so-called evidence and my own presuppositions? Do we therefore call something false because it is not in harmony with our preconceptions? Can something be false or right *per se*? What are the criteria used by people to accept some assertion as proven? My own answer to these questions is that I am a scientist who is able to state the conditions under which something happens that I claim is actually happening. I can supply arguments and reasons that meet the conditions of a scientific explanation, but what I am actually saying is neither true nor false.

POERKSEN: A proof or a scientific explanation is generally held to be a convincing and, in particular, absolutely valid form of evidence: a proof transforms an assumption or a hypothesis into a truth.

MATURANA: I would contest this. In my view, a proof is the acceptable offer of a description, which generates and produces the event that we want to prove. Proofs and explanations have nothing to do with the reflection of an external reality or truth; they are expressions of an interpersonal relation. We believe an argument or an explanation because we consider it valid, because it is described in a way that we – for whatever reasons, and on the basis of the most diverse criteria of validity – hold to be acceptable.

POERKSEN: Then the experience of truth is, in fact, a sort of experience of harmony?

MATURANA: Exactly. When the problems finally seem to be solved and the answers have been found, then all the doubting and searching is replaced by a state of contentedness; there are no more questions. Proofs and explanations fundamentally rest on their acceptance by individuals or groups. They change a relation. If we accept something, we always, consciously or subconsciously, apply a criterion of validation in order to decide about the acceptability of what is to be proved and explained.

THE EPISTEMOLOGY OF AN EXPERIMENT

POERKSEN: In your books you describe experiments with frogs, salamanders and pigeons. You studied perception in these animals; your epistemological insights are the products of your work in the laboratory. Do these studies merely illustrate the assumption that we can never know the real world, or is there more to them?

MATURANA: These experiments relate to my personal history and my experiences as a scientist; they must not be taken as evidence of truth; they illustrate the points of departure and the course of my own way of thinking. When I speak about the experiments with frogs, pigeons or salamanders, I refer to the circumstances in which my ideas developed at the time. In this way, the conditions are revealed that induced me to leave the traditional paths of perception research and to change the established system of epistemological questioning.

POERKSEN: Could you exemplify the history of your re-orientation by some relevant experiment?

MATURANA: Let me select a number of experiments carried out by the American biologist Roger Sperry in the forties. Roger Sperry removed one of the eyes of a salamander, severed the optic nerve, rotated it by 180 degrees, and carefully put it back into its socket. The optic nerve regenerated and the vision of the rotated eyes in the animals returned after some time. Everything healed but there was a crucial difference: The salamanders threw their tongue with a deviation of 180 degrees, when they wanted to catch a worm. This noticeable deviation corresponded exactly to the degree of rotation performed on the eyes; therefore, an individual animal turned round when there was a worm in front of it, and then threw out its tongue.

POERKSEN: What were these experiments intended to show or prove? What was the goal?

MATURANA: With these experiments, Roger Sperry wanted to find out whether the optic nerve was capable of regenerating and whether the fibres of the optic nerve would re-grow to join their original projection areas in the brain. The answer is: it does indeed happen. He also wanted to find out whether the salamanders are able to correct their behaviour, whether they are capable of learning – and whether they would, after repeated tonguing, hit the worm again in order to eat it. The answer here is: No, that is not possible; the animals keep tonguing with a deviation of 180 degrees; they starve to death if they are not fed. When I myself heard about these experiments and replicated them I realised, however, that Roger Sperry had formulated a misleading question that tended to obscure the observed phenomenon.

POERKSEN: In what respect was his research goal misleading?

MATURANA: He started out from the assumption that the salamander aims at a worm in the external world with his tongue. His question implied, as Gregory Bateson would have said, a whole epistemology, a specific worldview. It takes for granted implicitly that the external

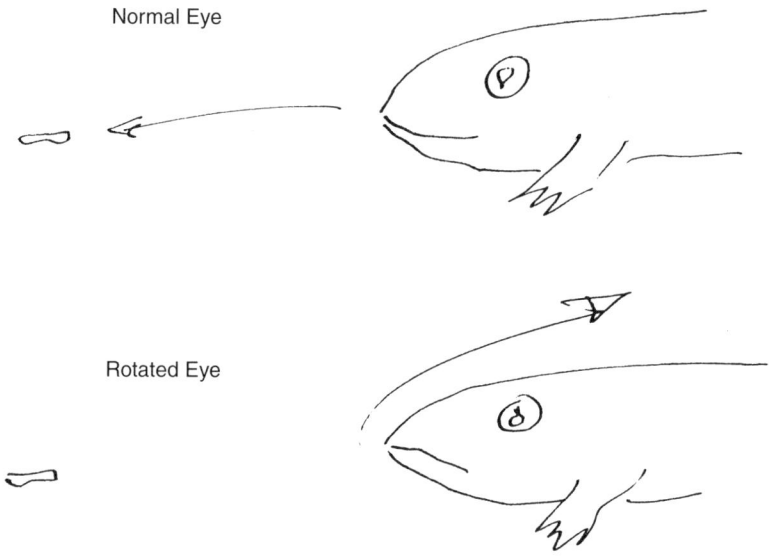

Fig. 4: The diagram shows two salamanders with a worm that has been placed in front of them by an observer. The salamander with the normal eye throws its tongue forward when the observer puts a worm in front of it. The salamander eats normally. The second salamander has had its eye rotated and throws its tongue backwards when the observer puts a worm in front of it. (Drawn by Humberto R. Maturana.)

object is processed in the brain of the salamander in the form of information about its shape and location. The salamander, consequently, makes a mistake; it does not process the information coming from outside correctly. However, I find it much more meaningful to interpret the experiment in a completely different way. The salamander, I would maintain, correlates the activities of its nervous system, which lead to the movement and ejection of its tongue, with the activities of a certain portion of its retina. If it is shown the image of a worm, it throws out its tongue; it does not, as it would appear to an external observer, aim at a worm in the external world. The correlation given in this case is an internal one. Seen in this way, it is not at all surprising that it does not change its behaviour, that it does not learn.

POERKSEN: In normal conditions, surely, there is a systematic connection between the world and its perception. If the salamander's eye is not operated on and changed round wilfully, the animal will catch the worm.

MATURANA: That is correct – and so we must ask ourselves how it is at all possible that a salamander with a nervous system that generates internal correlations, as a rule catches worms or other small animals with the utmost precision by throwing out its tongue. The experiment reveals some normality in the deviation; it makes us reflect on the conditions underlying that very normality. How does it come about that, in the usual circumstances, there is actually a worm in the precise spot hit by the tongue of the salamander? The explanation is given by the fact that the salamander and the worm normally share a common history and are part of an evolutionary process during which finely tuned relations of correspondence and mutual transformation have developed, a structural coupling between organism and medium. That an external observer is able to correlate features of the world outside (e.g. the presence of a worm) with the activities of an organism does not prove that the organism actually uses these features in order to orientate its behaviour accordingly.

POERKSEN: How did you yourself discover the hidden epistemology in the experiments of Roger Sperry? And what experiences and observations have helped you to develop the empirical epistemology that you are advocating today?

MATURANA: It was in 1955 in England that I replicated Roger Sperry's experiments – and it took another ten years for me to penetrate what I was actually doing and what had until then remained obscure for me. Only then did I understand the way the nervous system operated, i.e. its operation with internal correlations. When I performed experiments on the colour perception of pigeons in 1965 in Chile, I proceeded from assumptions quite similar to those made by Roger Sperry. My goal was to show how the colours in the external world, which I had specified in terms of their spectral composition to secure the replicability of my experiments, are correlated with the activities in the retina. I wanted to establish the connections between Red, Green and Blue and the activities of the retina and the retinal ganglion cells. What did the red, green, or blue objects release?

POERKSEN: So you thought likewise that the external object determines what happens inside the organism.

MATURANA: Quite right. I expected to be able to demonstrate an unambiguous correlation between the colour and the activities of the pigeon retina because I had already shown in comparable experiments that the activities of certain cells could, in fact, be correlated with specific shapes. I performed numerous experiments – nevertheless, I simply could not confirm the correlation I had predicted. It was impossible to find specific cells or cell groups that would react in a determinate manner to given spectral compositions.

WHY THE NERVOUS SYSTEM IS CLOSED

POERKSEN: So we face here an exemplary situation of all the researchers who are trying to test a hypothesis. They will usually carry on along the same lines, change their assumptions within the given framework, or develop a completely different and new hypothesis. What did you do?

MATURANA: At first I thought that my recordings were not yet accurate enough so I tried to refine them and to improve my recording instruments. My procedure was as follows: While the pigeons were shown colour tables, the activities of their retinal cells were recorded by means of fine electrodes. Numerous continually re-designed experiments only showed, however, that all the cells more or less reacted to all the different spectral compositions. No significant correlation between the activities of certain cells or cell groups and the spectral composition of the colours could be read off the minimally different modes of the cell reactions. The marginal differences in the modes of reaction were not significant.

POERKSEN: Comparing this experiment on the colour perception of pigeons with the striking behaviour of the manipulated salamanders, we find ourselves facing the same situation: the problem of the determination of what is inside by external determinants – coloured objects or moving worms.

MATURANA: That is the point. Furthermore, it becomes apparent that every experiment contains a particular view of the world, a complete epistemology, or cosmology, a bundle of expectations and premises, which guide our operations. One day, however, I realised that my expectations might possibly never be fulfilled because a correlation between the external stimulus and the internal reaction just could not be established. Only then I began really to appreciate Roger Sperry's experiments and their hidden epistemology, and to envisage the nervous system of an organism as a closed system. That was the turning point which gave my thinking a new direction.

POERKSEN: What precisely brought about that transformation of your views? What was the reason? You could simply have accepted the failure of the original hypothesis and moved on to a new topic.

MATURANA: And that is precisely what did not happen because I managed to bring about a re-orientation that demolished the framework within which a transformation would still have been acceptable. The customary way of slightly modifying one's assumptions and procedures would have consisted in creating ever-finer instruments of measurement and in continuing to carry out new experiments in the hope of producing productive results, in the end. I opted for something entirely new, however, which made some of my university colleagues seriously doubt my sanity. Perhaps I should, I said to myself, deal with the strange question whether the activity of the retina could be shown to be connected with the names of colours, which represent a certain experience; whether there might be an internal correlation between the activities of the retina and the names of colours, i.e. between different states of activity within the nervous system. The consequence was a momentous change with regard to the goal of my research and the traditional point of view. Suddenly I found myself outside the established traditions of perception research. Suddenly I was confronted by epistemological questions: What does it mean to know if we consider the nervous system as a closed system? How can the process of cognition be understood at all?

POERKSEN: Your key idea to correlate colour names and retinal states does indeed seem somewhat strange and rather curious. Names and designations of colours are, after all, merely arbitrary products of convention.

MATURANA: People naturally thought I was crazy. It even led to people laughing about me behind my back during my lectures when I had turned to the blackboard to write something. A friend of mine told me one day. I knew very well, of course, that names are arbitrary entities; at the same time, I was aware of the fact that we use the same colour term for extremely diverse spectral compositions; our colour terms, therefore, refer to our own experiences, they are indicators of experiences. What had to be demonstrated was that the activities of the retina and the retinal ganglion cells are correlated with the specific experiences represented by colour names. That is precisely what I managed to show.

POERKSEN: What, then, is a colour?

MATURANA: It is nothing external but something happening in an organism – merely released by an external source of light. A colour is what we see, what we experience. The colour designation refers to the particular experience of an individual in certain situations, which is independent from the given spectral composition of light. My approach was to compare the activity of the nervous system with the activity of the nervous system, to relate the activity of the nervous system to itself, and to view it as a closed system. My focus was on internal correlations.

POERKSEN: Even this version sounds strange and obscure, at first. The classical view, after all, defines the nervous system of an organism as an open system: receptors react to excitation by external stimuli, and these are then processed further. The result is a more or less faithful representation of the real world.

MATURANA: Those who share my conception and accept it as a basis for their own reflections first have to get rid of an erroneous interpretation of the concept of information processing which was once quite popular in biology but did not really contribute very much to our understanding of the nervous system. For a long time, the dominant belief was that the nervous system of an organism processes information coming from outside in order to generate adequate behaviour of this organism. The source of information located in the environment would – so the assumption – modify the structure of the organism in such a way as to generate behaviour that would be adequate with regard to given external circumstances. Such a view, however, is not helpful at all; the nervous system does not function in this way.

POERKSEN: How would you describe what is going on? What happens, in your view?

MATURANA: When light reflected by an object that the observer describes as external reaches the retina, an activity is initiated that is enclosed in the structure of the retina itself (and not in the structure of the source of light, nor in the structure of the world). The external world can only trigger such changes in the nervous system of an organism as are determined by the structure of the nervous system itself. The consequence is that there is no possible way, in principle, for the external world to communicate itself in its primordial, true form to the nervous system.

POERKSEN: What does this mean? How does this abandonment of the idea of information processing inspire or even compel us to think and speak differently about the external world, the organism, and the nervous system?

MATURANA: Our approach changes completely. We can no longer accept descriptions of the nervous system as a system that computes representations of an external world and processes information coming in from outside, which then results in adequate behaviour and appropriate reactions of the organism. The nervous system now appears as a structure-determined system with its own specific mode of operation. Any change in it is only triggered but neither determined nor specified exclusively by the features and properties of the external world. It computes nothing but its own transformations from state to state. People who accept this insight must draw a strict conceptual distinction between the operations taking place inside the nervous system and all the processes occurring outside it. They must also be quite clear about the fact that there is no inside and no outside for the nervous system but only a perpetual dance of internal correlations in a closed network of interacting elements; inside and outside exist for the observer but not for the system itself.

Double look

POERKSEN: It seems to me that such an interpretation of neuronal processes must inevitably lead to the biologically grounded denial of the external world. What you are saying raises once more the suspicion of solipsism. The nervous system exists, if I understand you correctly, in complete cognitive isolation. It is floating along as if in a void.

MATURANA: Once again I have to reject this classification of my views as solipsistic. Again: as the observer that I am, I do not at all deny the experience of an external world, nor the experience of common discourse and the experience of other people's existence. I vehemently deny, however, that the operations of the nervous system can be related to this external world and its features in any meaningful way or that they can be derived from them. The nervous system operates as a closed system of changing relations between neuronal states of activity that continually lead to other changing relations between neuronal states of activity. For its operation as a system, nothing else exists but its own internal states. Only observers can distinguish between inside and outside or input and output and can, consequently, diagnose the impact of an external stimulus on internal processes and the organism or, conversely, an impact of the organism on the external world. What is described as adequate behaviour is the result of a relation established by observers. They have related the features of an external world to the organism and its nervous system although these external features are not part of the operations of the organism nor of its nervous system.

POERKSEN: Surely people who speak of the closure of a system can neglect the existence of an external world, can challenge and deny it.

MATURANA: The assumption of closure has to do with the internal dynamics of the nervous system and refers to its mode of operation; it has nothing to do with the question whether there is – independently from the closure of the system – an external world or whether we must consider reality an illusion. That is no longer the problem. Once we have accepted that there is no possibility of making testable claims about an observer-independent reality, the fundamental change in our epistemology has been completed. All forms of observation and explanation are now expressions of the system's operations with whose production we may now deal. A re-orientation has come about, a change from Being to Doing, a transformation of the classic philosophical questions.

Poerksen: The assumptions of the closure of the nervous system and the external viewpoint of the observer imply, if I understand correctly, the distinction of two perspectives of observation. On the one hand, observers describe external impingements upon a system and construct correlations between stimulus and response, input and output, cause and effect. On the other hand, the system operates – independently of external influences – in its own specific manner.

Maturana: That is it. The phenomenal domain of physiology and internal system dynamics does not intersect with the domain of behaviour and observable movements in an environment. These domains cannot be reduced to each other, nor can the phenomena of one of the domains be derived from the other.

Poerksen: Could you provide an example?

Maturana: On certain occasions I use the example of an instrument flight when I want to explain the difference between the internal operational dynamics of a system and what happens in the domain of interactions of the system as a whole. Imagine pilots sitting in the cockpit and flying a plane in complete darkness. They have no immediate access to the external world nor do they need it, they act on the basis of measurement values and indicators, employing their instruments when the values change or particular combinations of values emerge. They establish sensorimotor correlations in order to keep the relevant values within specified limits. When the plane has landed, friends and colleagues may appear who have observed the plane arrive, and congratulate the pilots on their successful and admirable landing in thick fog and dangerous storms. The pilots are confused and ask: "What storm? What fog? What are you talking about? We just handled our instruments!" You see: What happened outside the plane was irrelevant and without meaning to the operational dynamics inside the plane.

Poerksen: Do you want to suggest with this pilot example that we are all enclosed in our own cockpits and our worlds? More drastically: Are we in the same situation as these pilots when we are trying to understand the world? If this should be so, then I would maintain that we would be incapable of diagnosing our situation the way the pilots do because we could not possibly know the limits of our knowledge. If we were able to do that, the limits would cease to be limits.

Maturana: Correct. Only one condition permits us to perceive our own blindness: We must be able to see and to know, and we must, therefore, no longer be blind when we gain insight into our own blindness. However, this is not the point of the example. For the pilots just working their instruments in the situation described, the so-called limits of knowledge do not exist at all. The crucial thing is that only observers can speak of limits because they have access to their own domains and to the domain of the internal operational dynamics of the cockpit. They have to use their double look and compare what happens inside the cockpit with the circumstances in the external world and then relate what they have seen in the two different domains in another domain generated by themselves. All that observers can say is the result of this double look.

Poerksen: These observers describing the perceptual limits of the isolated pilots must essentially be realists. They are able to grasp the reality that remains unknown to the pilots in their cockpits, and therefore they know what really happens.

Maturana: How can these observers know that they themselves are not sitting in some cockpit containing a world in which pilots sit in cockpits that may be observed through double look? They could only diagnose limits of knowledge if they had absolute knowledge about precisely that situation. Under this condition only they would be able to establish the limits of knowing, with the necessary consequence that they would have to declare themselves realists believing in some objective realities. I would claim, however, that these observers are comparing two different domains of distinction, not a real world with a merely fabricated world. They see the pilots at work inside, as if through a peephole in the wall of the plane, whereas from outside they see the plane as a whole in relation to its domain of operation.

POERKSEN: You say that the thesis that the nervous system is an open system results from a particular perspective chosen by observers. But is your claim that the nervous system is closed and cannot meaningfully be described in terms of input and output not the result of an observer's chosen point of view? Surely, both these assumptions cannot be right at the same time. They are essentially contradictory.

MATURANA: There are, in fact, two different perspectives of observation that naturally generate different descriptions. Still, the two conceptions are not equally valid. If one wants to find out how the nervous system operates on the assumption that it is an open system, one has chosen a misguided approach. Observers will believe, accordingly, that its mode of operation is dependent on its input. What they define as external stimuli in the environment will gain enormous importance and make them overlook the internal dynamics of the system and to confuse the domain of their descriptions with the domain of the internal dynamics of the system. Such a confusion of domains cannot offer an adequate explanation of the mode of operation of the nervous system. – If we, however, view the nervous system as a closed network we can understand its mode of operation and recognise how structural changes in an organism that is in correspondence with its medium will lead to structural changes in its nervous system and ultimately to changes in its behaviour. We need, therefore, no longer speak of the flow of information but instead, when observing an organism in its environment, ask ourselves how the strange structural coupling between the activities of the nervous system, the body of the organism, and the external circumstances functions in detail.

POERKSEN: What does it mean, then, to conceive of the nervous system as a closed system? It cannot be totally shut off against the environment because it is dependent on the exchange of matter and energy. If this exchange fails for some reason, the organism will collapse and perish. So the input from outside cannot simply be neglected; every living being is vitally dependent on it.

MATURANA: Now you are arguing like a physicist in the context of thermodynamics. Naturally, the nervous system of an organism must be open for the flow of energy and matter – that is more than obvious. The cells will die otherwise. Closure is not a physical concept but characterises the self-referential working of an internal dynamics. The processes recurring in a particular domain remain in this domain; we are dealing with the operations of a system that determine its boundaries and make it a determinate entity. Therefore, by closure of the nervous system I mean that its states of activity always lead to other states of activity and are triggered by states of activity, and that all these diverse states of activity remain within the network of neuronal elements.

TO LIVE IS TO KNOW

POERKSEN: You have already reported how certain intellectual experiences completely changed your epistemological views. The question now is how we can understand and describe the processes of cognition if the nervous system is considered as a closed network operating exclusively according to its own internal laws. What is cognition?

MATURANA: Cognition is the observation of adequate behaviour in a particular domain, not the representation of an independently existing reality, nor a process of computing according to the conditions of the environment. When an animal or a human being behave adequately and are in coherence with their circumstances, and when observers come to the conclusion that there is adequate behaviour in the situation they observe, then these observers will say that the animal and the human being in question possess knowledge, that they manifest cognition. Knowledge is, in other words, behaviour in a particular domain, which is judged adequate by observers.

PoerkseN: Your description of the circularity of the cognitive processes leads to a circular definition of cognition and knowledge, which mirrors the whole architecture of your theory. Cognition is understood and established by observers; knowledge appears as an observer-dependent product, not as something objective.

Maturana: This is the idea, quite. Observers interpret the interaction of organisms with their environments in such a way as to diagnose adequate behaviour, and observers attribute knowledge to the observed systems and evaluate their actions as indicating cognitive operations because they consider them adequate and appropriate. The maintenance of life is an expression of knowledge, in this sense, a manifestation of adequate behaviour in the domain of existence. In the form of an aphorism: *In the living of living beings living entails knowing, and knowing entails living.*

4. On the autonomy of systems

THE LIMITS OF EXTERNAL DETERMINATION

POERKSEN: In the process of your epistemological re-orientation you learned from experiments. This is the classic procedure of the realists: They test a hypothesis, it fails, – and they modify it. The circumstances, the real world, force them to revise their ideas. The course and the direction of your thinking, are they not essentially realistic?

MATURANA: This is an interesting point. We might, of course, say that I acted like a realist when I changed the traditional problems of the theory of knowledge in such a way as to become an opponent of realism. But that is not of primary importance. I would claim that it was as a scientist and not as a philosopher that I tackled the problem of the possible existence and the degree of influence of an external reality. The distinction between science and philosophy that I am suggesting here has to do with the question of what the philosopher and the scientist want to preserve when they develop a theory. Their intentions are different.

POERKSEN: What are they like? Could you clarify your distinction between philosophy and science in detail?

MATURANA: Philosophical theories arise when we try to preserve certain explanatory principles that we consider valid *a priori*. This interest in the preservation of principles and their coherence justifies disregarding what may be experienced. Scientific theories, on the contrary, arise when we want to preserve the coherences in relation to what we are capable of experiencing. The scientist can, therefore, ignore principles, dissolve them, – and design a scientific theory.

That is precisely what I did. I began with the coherences within experience, I investigated the colour perception of pigeons, i.e. I investigated the operations of living systems – and had to do terrible things to them for the purposes of my research. The question as to whether an external reality really existed had little relevance for me; it was not one of my problems.

POERKSEN: Can you see experiments and experiences that might refute your present claims and put you back on the path of realism?

MATURANA: I could only give up my views if the structural determinism to which all systems are subject were no longer in force. What happens in any system, we must bear in mind, is necessarily determined by its structure and not specifiable by external influences.

POERKSEN: How would you like such a theory to be understood? What kind of truth status does it have? Is it perhaps even true in an emphatic sense?

MATURANA: Of course not. The assumption that living systems are structure-determined systems is in no way related to an observer-independent reality; it is an abstraction resulting from the coherences that observers may experience. To abstract means to grasp the regularity of some process and formulate it without paying attention to the actual elements involved. Whenever I discuss the structural determinism of a system, I do not describe ontic or ontological facts or some truth, I merely present an abstraction from my experiences as an observer.

POERKSEN: What do you mean by *structural determinism*? How would you define the concept?

MATURANA: When you press – for instance – the key of your tape recorder with your index finger in order to record our conversation, then you expect the machine to record. Should the machine fail to do so, you would certainly not go and see a doctor to have the functioning of your index finger checked. You will take the tape recorder to someone who understands its structure and will, therefore, be able to repair it so that it will react to the pressure of your index finger in the appropriate way. This means that we treat your little tape recorder as a system in which everything that happens in it or to it, happens determined in its structure. I call this condition structural determinism, and I call this kind of system a structure-determined system. Moreover, we human beings deal only with structure-determined systems, and we are structure-determined systems ourselves.

POERKSEN: In what ways? Could you give another example?

MATURANA: Suppose you see a doctor about a pain in your stomach. You will be properly examined – and perhaps your appendix will be removed. So you will be treated like a structure-determined system: the pain you felt before the operation and the relief you experienced afterwards were both determined by your structure and its modification by the doctors. More generally, this means that an external agent impinging on some molecular system triggers certain effects but cannot determine them. Any impingement from outside merely triggers some structural dynamics; all its consequences are, however, specified and determined by the structure of the system itself.

POERKSEN: Is this so? Let us assume I offer you medicinal tablets or hard drugs and we both take some; we shall experience similar things. Drugs have quite specific effects.

MATURANA: Perfectly correct, but the similarity of our experiences does not refute structural determinism at all. Taking drugs means bringing molecules with a specific structure into your organism, which then become part of it and modify the structure of its nervous system. What happens will, however, necessarily depend on the structure of the nervous system itself. Without receptors inside the organism for the substances you put in, nothing can happen at all. A receptor, one must remember, is a specific molecular configuration that matches the structure of the substance in question, a drug, for instance. In this way, a change in the organism is triggered.

ORGANISATION AND STRUCTURE

POERKSEN: Perhaps we should put aside, for the moment, further illustrative examples and turn to the vital problem of the new concepts and the new kind of language that are definitely needed now to express what is traditionally called *stimulus* and *input* and is supposed to control the behaviour of living beings. Terms of this kind, although still widely current in our everyday thinking, can no longer be used because they obviously imply direct and monocausal influence.

MATURANA: That is correct. The mistaken concept of instructive interaction must be corrected by an alternative idea: Whatever happens in a living being is determined by its structure and not by the structure of the external agent. From the perspective of the commenting observer, I speak, therefore, of *perturbations* to which a living being is subjected. The observer perceives some entity that, in his view, impinges on the system and triggers structural changes in it that do not lead to the destruction of the system but permit it to preserve its organisation. This form of encounter I call perturbation. Another possibility is that the system loses its identity and falls apart. In this case, a destructive change has taken place. When somebody pushes me I can say: Don't perturb me! When she hits me on the head with a hammer, however, the potential change of my structure may be dangerous and lead to my destruction. The correct expression for me to use would be: Don't destroy me!

POERKSEN: Could you describe these variants of change in people, things, and systems more precisely?

MATURANA: Here is a little story. One day I gave one of my sons a number of tools but forgot to give him wood to practise a little carpentry and test his new tools. When I came home from work, he had sawn off a corner of our table to have some wood for his purposes. "You have," I said to him, "modified the structure of my table." The table could still be used and had not lost its identity. Its structure was different now; its organisation had stayed the same. A few months later, looking for a board, my son had sawn a large chunk out of the tabletop. I could explain to him now that he had not only changed the

Fig. 5: A table which has undergone a structural change with conservation of organisation. (Drawn by Humberto R. Maturana)

structure of the table but also destroyed its organisation. "Now," I said to him, "I have no longer a table." What the story tells us is that distinguishing between the organisation and the structure of a system allows us to specify more precisely how a system may change. Had I wanted to make sure that the table remained intact I should have explained this to my son early enough.

POERKSEN: This conceptualisation of yours solves the classic problem of identity and change, stability and transformation. It answers the old question of philosophy: How can something change and yet remain the same?

MATURANA: The distinction between structure and organisation permits us to grasp the different ways in which any system may change

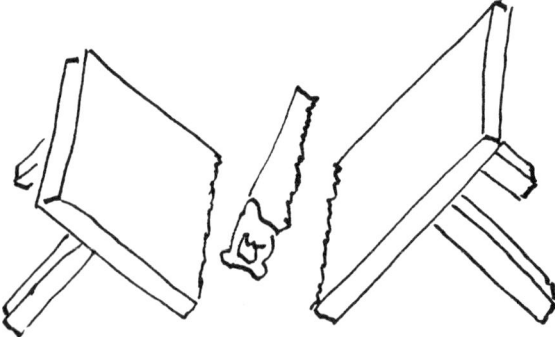

Fig. 6: A table which has undergone a structural change without conservation of organisation. The table is no longer a table. (Drawn by Humberto R. Maturana)

and remain recognisably the same system. We can alternate flexibly between the consideration of identity and change. The structure of a system, which may change and whose modification may lead either to the preservation or the destruction of the organisation of the system, refers to the components actually given and the relations between these components that constitute a composite unity as a special kind of unity. The structure of a unity makes this unity a singular case from a particular class of unities. A table may have quite diverse structures; it may, for example, consist of wood, glass, metal, or some other material, but this does not affect its identity as a table. The organisation of something is, however, invariant. It refers to the relations between the components that let us recognise what class a composite unity or system belongs to. A table is – independent of its particular structure – always recognisable as a table because it exhibits a particular organisation. As my son demonstrated, the structure of a table may be changed so drastically that its organisation is destroyed, too; the table no longer exists, having lost its "tablehood."

POERKSEN: How should we evaluate the sort of structural change that you call perturbation? The concept of perturbation has often been paraphrased as "disturbance" or "interference," making the environment a mere source of irritation for living beings, e.g. humans. This sounds very negative. I would prefer to think that perturbations might just as well be inspiring and uplifting events.

MATURANA: Certainly. A perturbed person may be inspired, perhaps irritated, even disturbed, or terrified. Any system-independent evaluation of a perturbation, negative or positive, would be misleading. The concept cannot be used to justify any such evaluations.

POERKSEN: Can the distinction between the traditional idea of input and the concept of perturbation be made more precise? What is the central difference?

MATURANA: The concept of input implies that there is direct influence, that something from the external world enters the system and determines what happens and occurs there. Such a view simply cannot be defended because it rests on the false presupposition of instructive interaction and contradicts the structure-determinism of all systems. When a perturbation occurs, a system encounters an entity that triggers a structural change without destroying the system. The concept of perturbation is in correspondence with the idea of structural determinism.

POERKSEN: We could say, though, that those who are unable to intervene instructively and manipulate in a direct way simply do not know enough. They do *not yet* understand the systems in question. Apparently all the gurus, the psychotechnologists, and the successful salespeople, possess sufficient knowledge to be able to control the behaviour of living systems – other people – in a very efficient way? Seen in this way, the impossibility of instructive interaction is a problem of knowledge and of the difficulties of understanding.

MATURANA: Of course, people may believe that they have special abilities and insights and are therefore able to transform a perturbation into an input and perform an instructive interaction after all. Such an erroneous conviction cannot, however, in any way serve as an argument to invalidate the structure-determinedness of a system of whatever kind. Two systems can encounter each other only on the level of their structures; and their specific structures – the components and the relations between these components – determine what happens in each system due to this encounter. When we analyse what the gurus and the successful salespeople actually do in the course of their manipulative activities, we realise immediately that they always operate with a special understanding of the structures of the systems which they perturb. They exploit the properties of the systems, e.g. the character traits of humans, their desires and needs, and with their insights they are able to trigger some behaviour in the other people which serves their own interests.

POERKSEN: Is such insight not dangerous? If one has a grip on the logic of a system, the idea of manipulation is not far away: Systemic insight becomes the basis of even more effective control and dominance.

MATURANA: I do not share such a view. I think that people who understand a system and use their knowledge accordingly, need not necessarily act in a manipulative way; such an evaluation of their actions requires knowledge of the emotions underlying these actions. Actions based on the understanding of a system might, on the contrary, be interpreted as an expression of particular wisdom. This means that I do not consider manipulation as a specific kind of action but rather as a specific emotion giving shape to some special activity. To manipulate means pretending to do something for someone but actually operating only in one's own interests. Manipulating people means cheating and lying. A liar knows that he is lying. That is, if you wish, the beauty of a lie.

POERKSEN: If I had to reduce our conversation about structural determinism to one conclusion, it would be this: systems are autonomous; one can invade them only according to their own specific conditions but not determine what occurs and happens inside them. Would you agree?

MATURANA: I would agree as long as autonomy is understood as *self-governance* and does not imply that a system can be separated from its medium. That would be completely unthinkable. There is no autonomy in this sense because every living system exists in a medium. What influences a system, however, is determined by its internal dynamics, which shapes these influences in quite particular ways. When the system finally dies, this means that it was incapable of keeping itself alive, that it lost its autonomy.

UNDERSTANDING RESPONSIBILITY

POERKSEN: In what sense are human beings autonomous? It would certainly not be quite correct to say that they are completely free.

MATURANA: Autonomy in the human domain means that what is uniquely characteristic of a person is preserved. Freedom is something else: a human experience requiring reflection. Strictly speaking, there is no freedom at all; strictly speaking, there are no alternatives because every happening and every action results from the correspondence with the structural coherences of the moment. People who are ignorant of the given structural coherences believe that they see alternative ways of action. Arriving at a road junction, they can choose between two directions. They see, for instance, two alternatives for continuing the journey, which they consider identical because they do not know which to take, which one is better. In such a situation, they must first create a difference and learn to see both directions as distinct in order to be able to choose. Perhaps they will flip a coin and in so doing make way for processes revealing a difference that will finally permit a decision in correspondence with the given structural coherences of the moment.

Poerksen: You insist that human beings are structure-determined systems, too; they are autonomous but not free. Stressing the feature of determinism in such strong terms, how can you still speak about responsibility in a meaningful way? My thesis is: Only those who recognise themselves as free can claim responsibility for their actions.

Maturana: Perfectly correct. Living systems cannot act responsibly because they know no purpose or goal; they simply live in the flow of existence. Only human beings can assume responsibility in the domain of relations because they exist in language. They are capable of describing a certain action as responsible. Language enables us to reflect and distinguish the consequences of our actions for other living beings. In this way, our caring for other people gains presence – and the possibility of responsible action arises.

Poerksen: But, surely, this requires freedom. Any person desiring to act ethically must have the freedom of choice and self-determined decision. Repeating the question: Do not your key concepts of structural determinism and your special understanding of autonomy force you to abandon the idea of freedom and, consequently, the possibility of responsible action?

Maturana: The experience of choice and decision, which we human beings make, does not at all contradict our structure-determinedness. Human beings will always remain structure-determined systems; they may, however, by virtue of a perspective opening up in a meta-domain, make the experience that they have a choice. Then they move in another domain but still operate as structure-determined systems. This experience of the potential choice between different possibilities, however, is a unique characteristic of the human species and requires language. Having a choice presupposes the ability to observe and compare at least two different situations appearing at the same time, and then to adapt one's perspective in such a way as to be able to make out a difference between these situations. At first one sees only sameness and is blocked. A change of perspective and position may enable us to see potential distinctions in what appears to be the same; then we can move – according to our own preferences and ways of life – and choose one possibility while negating others. As this process is an intentional act in the language of living beings, it is possible to classify it, from the point of view of an observer, as a process of choice.

POERKSEN: Does this mean that it is the meta-perspective that makes it possible to identify an action as an act of choice and decision?

MATURANA: Exactly so, yes. Only from this perspective does it become possible to characterise something as a choice and a decision between different possibilities. We perform an operation on a meta-level because we have the ability to use language and to make ourselves aware of an event and its consequences. In this act of becoming aware, the phenomena we are dealing with are transformed into objects of contemplation. We gain a form of distance that we lack when we are completely immersed in our activities and situations. If we accept this and consider it adequate, an action may then be described as *responsible* or as *irresponsible*.

POERKSEN: Could you use an example to illustrate this?

MATURANA: Some time ago, reports travelled round the world that a boy who had been trying to get to Miami together with his mother in a small boat from Cuba was saved from drowning by dolphins. For some reason, their boat sank and the mother drowned. The boy, however, was kept afloat by a school of dolphins, saved from drowning, and finally rescued. What those dolphins did we can, as beings living in language, describe as *responsible*. The dolphins do not, as far as we know, possess the ability to comment on their activities and to tell us about what happened between them and the boy floating on the sea. However, *we* are capable of talking about the relationship between those animals and the boy because we operate in the domain of language. We can characterise what happened as an effort to keep another being alive. From this meta-perspective the activity of the dolphins appears as a responsible action.

POERKSEN: To act responsibly, then, means to care for someone else and, at the same time, to reflect on the consequences of what one is doing in relation to the circumstances in which one does what one is doing.

MATURANA: Exactly. People are aware of the circumstances and reflect the consequences of their activities. They ask themselves whether they want to be what they are as they are doing what they are doing. In the moment of self-observation, all the certainties and securities of the state without reflection disappear. When, through the linguistic operation, a form of contemplation and an awareness has been generated that allows observation, then people will, at the next step, act according to their own preferences, that means they will act responsibly. And when they, at a further step, try to find out whether they value their own preferences and intend to maintain them, then they are free. Do I like my predilections? Do I like the decision I have taken and which I have just said I like and that it corresponds with my desires? In this moment of the reflection of their own choice, there arises the experience of freedom, although they nevertheless operate as structure-determined systems.

POERKSEN: I want to keep on questioning: How can a structure-determined system feel responsible for its own actions? If I cannot control and influence others, then the effects of my activities become completely incalculable. We are confronted by a *paradox of responsibility* because we are to be held responsible for something the consequences of which we could not possibly foresee. Doing good may potentially trigger terrible consequences (and vice versa).

MATURANA: The concept of responsibility is ambiguous. Some authors mean by responsibility that we must be accountable for all the possible consequences of an action. Responsibility then means causation. For me, responsible action is a question of awareness. Persons act or fail to act in the awareness of all the possible and desirable consequences of their actions. It is not necessary for the consequences of an action to be fully calculable and foreseeable; there may indeed be undesirable consequences in the end. In my view, being responsible simply means being in a certain state of attention and mindfulness: one's activities and one's desires correspond in a reflected way, that is all.

Poerksen: The concept of responsibility is, for you, not linked to the idea that it is possible to plan the consequences of an action?

Maturana: This is not relevant. To plan something means to envisage ways and procedures for achieving a certain result and to subordinate the next chosen steps to this imagined result. These consequences need not come about, however, and perhaps they exist only in the minds of particular people. It is crucial, in any case, that the people designing things in this way live responsibly and act in full awareness of the possible consequences of their actions. They are responsible for what they say and do. Nevertheless, they are not accountable for what other people make of what they say and do.

A miracle is needed

Poerksen: You locate the experience of responsible action and freedom at the level of reflection. In this way, as I see it, the experience of freedom may be reconciled with structural determinism. How can we now view the phenomenon of surprise from your perspective? The idea of the structure-determined system certainly suggests that all behaviour can be calculated and predicted.

Maturana: People making predictions speak of their expectations as observers. They believe that they know all the factors influencing a system and assert that certain states will result from other states that we can then observe. Living systems, however, are not calculable in this sense although they operate in a structure-determined way. Structural determinism does not entail predictability but is related exclusively to the structural coherences of the moment, which change all the time. By the structure of a system I mean, let me recall, the components and the relations between these components, which make it a particular kind of system. When the components or their relations change, the structure is transformed. If you slide around on your chair, then you change your structure; if you speak or keep silent and listen, then your structure changes. It is not rigid and firm but in constant change.

Poerksen: We are now left with the mental exercise of exploring under what conditions structural determinism might no longer be operative. In other words, can you state conditions under which something dead or alive would no longer be subject to universal structural determinism?

Maturana: Only the advent of a miracle can violate structural determinism. Suddenly the impossible seems possible, inexplicable and totally unexpected things happen. An example: Imagine a person that deserves to be called a saint. Proof of the person's sainthood are frequent miracles supposedly worked by, and attributable to, that person. There are people suffering from diseases which are incurable according to present medical knowledge who, having prayed intensively to the saintly person for help, suddenly get well again to the doctors' surprise. Their affliction disappears and they recover again. What has happened? We do not know, and it will perhaps remain a mystery forever. A phenomenon of this kind exhibits the central property of a miracle: the apparent suspension of structural determinism.

Poerksen: The philosopher Karl Popper and the disciples upholding his theory of science demand that the conditions within which assumptions may be tested must always be stated explicitly in order for them to be refuted or falsified. Only the fulfilment of this requirement can raise an assumption to the status of a proper scientific hypothesis. Does it not give you an uncomfortable feeling to have to accept that structural determinism is more or less unfalsifiable? A singular miracle experienced by a few people cannot really serve as a counterexample.

Maturana: Please remember that Karl Popper merely wants to define the particular situation or the specific phenomenon that might potentially falsify a hypothesis. We must be able to envisage the conditions of falsification: that is the requirement he proposes. And it is precisely this requirement that I can meet by stating the crucial condition of falsification: Only a miracle could invalidate structural determinism. The practical difficulty or impossibility of falsification is not relevant in Karl Popper's theory about how to decide whether some assumption is a scientific hypothesis or explanation. An explanation remains valid until it is refuted.

Poerksen: Are you expecting a falsification? Are you waiting for a miracle?

Maturana: No. And I do not think that we could really do very much with miracles; they seem to me to be rather impractical events. Just remember the story of King Midas of Phrygia who offered his services to the god Dionysos. It shows – in a satirical way for me – the uselessness of miracles that suspend structural determinism. Dionysos asked King Midas what kind of reward he wanted for his services. King Midas answered that he wanted everything he touched to turn into gold. And that is what it happened. He touched the grass – it became gold; he touched the table – gold! Happily, he went home, and his daughter came running towards him; he embraced her – and she became rigid and turned into a golden statue. What is the tragedy of King Midas? My answer: His tragedy was that he had no chance of becoming an analytical chemist. Everything he touched was the same for him: gold.

5. How closed systems interact

IMPROBABLE INTERACTIONS

POERKSEN: Professor Maturana, for a week now we have been meeting day after day for the purpose of this interview. Sometimes we sit together in your house, then we see each other in the rooms of the University of Santiago de Chile, and frequently our appointments take place in the institute you have just established in the city centre. What precisely is happening here? In the terminology you have introduced so far, we would have to say: a single structure-determined observer with a closed nervous system encounters another structure-determined observer with a closed nervous system. How can that be? How can two closed systems – an epistemologist from Chile and a journalist from Germany – meet for an interview in this mega-city of Santiago? Why do we not miss each other most of the time? Why are we apparently successful after all?

MATURANA: The reason is that our encounters take place in a domain of interactions that must be clearly distinguished from the operational domain of our nervous system. When we make appointments and meet, we act as organisms, as wholes in a sphere of relations. Our meetings do not take place on the level of the internal operations of the nervous system; that is certainly not the place where we meet.

POERKSEN: Nevertheless, we have so far been talking exclusively about *lonely systems*. Therefore, the thought suggests itself that we cannot but permanently misunderstand each other, that we ought to be permanently annoyed by each other's self-directed, autonomous behaviour. This simply is not the case, it does not happen. How is it possible to transcend the loneliness? How can the two of us – closed systems – converse and even attempt to compile a book together?

MATURANA: As human beings, and as the mammals that we are, we share the property of enjoying the company of others, conversations and communal action, – and so we keep returning to these enjoyable forms of communality in our everyday life. The fact that we are closed systems is irrelevant in the domain of interactions; we remain lonely inside but together we create a domain in which our encounters take place. Our conversations take place in the flow of interactions, in a domain that must be distinguished from what goes on inside us.

POERKSEN: We are, as you claim, closed systems, and exist in a sphere of unbridgeable loneliness, but at the same time we also meet with each other and make plans together. How does that fit together? The two positions flatly contradict each other.

MATURANA: No, they do not. Faulty thinking causes the contradiction you suspect here. The mistake is to confuse two domains that must be kept distinct, and to try erroneously to connect what happens inside the nervous system with the events in the domain of social relations. That cannot work because each of the two domains must be considered separately. Therefore, the closure of the nervous system and the fact that we are able to make appointments do not contradict each other at all.

POERKSEN: I cannot follow. To arrange a successful appointment, surely the originally closed system must open up, switch to reception, make itself permeable, and engage in resonance. Everything will fail if it remains closed.

MATURANA: Here is a little analogy. Imagine you buy a new pair of shoes and begin wearing them from time to time. A year later, your feet and your shoes will inevitably have changed. They are no longer the same. The shoes have become much more comfortable although they have not in any way mingled with your feet; shoes and feet still exist as separate and closed entities. Their boundaries are clearly recognisable and have not become permeable for each other in any way.

The comfortable feeling due to the continual use of the shoes is not the result of an opening of the two distinct systems; it simply arises in a different domain.

POERKSEN: If we continue with this analogy, how could the interaction be described more precisely?

MATURANA: The central point is that foot and shoe, to stay with this ordinary example, both have a plastic, variable, structure. It transforms itself depending on the recurring and recursive interactions – and therefore foot and shoe can change together and in mutual correspondence in the course of time. The degree of congruence increases. This mutual change requires, however, that the shoes are used with certain regularity and frequency, and that there is a certain comfortable feeling that invites us to put them on more often. I claim that we can describe the encounters between human beings and other animals in the same way as the interaction between feet and shoes. The congruent changes – that is the whole secret – are the simple result of recurrent and recursive interactions between systems. These interactions trigger mutual structural changes that do not affect the organisation of the systems.

POERKSEN: What we have now is a theory of interaction, which does not contradict the fundamental autonomy of the systems and necessarily excludes any kind of reductionism. Keeping different domains and the phenomena arising in them distinct makes it impossible, if I understand you correctly, to play the game of reductionism – *this* is really nothing else but *that*.

MATURANA: Exactly. And suddenly it is possible to perceive phenomena which do not take place inside a system but in the domain of its relations, although they are, of course, in no way independent from the internal features of the interacting systems. Just look at the microphone through which our conversations are recorded. It is standing on the table on the tablecloth. When you pack it away this evening, we will both notice a slight indentation in the tablecloth due to an interaction. The indentation in the cloth is neither an internal feature of the microphone nor of the cloth but is certainly dependent on the characteristics of both – and belongs to the domain of their relations.

If we apply this to living systems, we can say: The nervous system and the whole organism may be closed, but if they have a plastic structure that changes in the course of the interactions they undergo, then a history of relations may unfold that does not intersect with the internal dynamics of the nervous system or the organism (and *vice versa*).

STRUCTURAL COUPLING

POERKSEN: How would you describe in your language what happens between us? What happens when we meet, talk to each other, make further appointments, and then continue with our discussions?

MATURANA: In my terminology I would say that the recurrent and recursive interactions generate *structural coupling*. With this term I want to refer to the history of mutual structural changes that makes it possible for a consensual domain to emerge, a behavioural domain of interlocking and reciprocal interactions between two structurally plastic organisms. With regard to our interview: We keep meeting and, therefore, are not only in recurrent, constantly repeated interaction but also in recursive interaction. Our conversations form the basis for further conversations, the elements of our conversations refer to themselves and build on each other, – and that is recursion. Our meetings trigger structural changes inside each one of us, and they continue as long as we move in the dynamic congruence that leads to structural coupling. Structural coupling arises if the structures of two structurally plastic systems change through continual interaction without destroying the identity of the interacting systems. In the flow of such coupling, a consensual domain is formed: a behavioural domain within which we act together and in reciprocal correspondence; the changes of state of the coupled systems – more generally – are reciprocally conditioned through interlocking sequences.

POERKSEN: These three concepts, – *recursive and recurrent interaction*, *structural coupling* and *consensual domain* – contain answers and problems. However: What problem do they solve? What question do they answer?

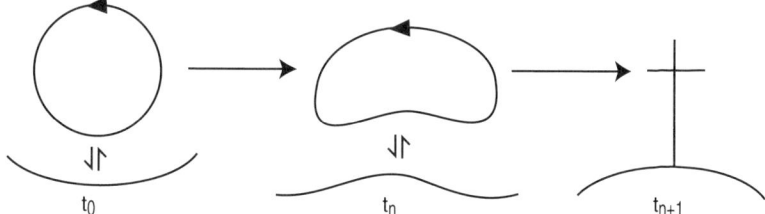

Fig. 7: This diagram shows a living system changing in interactions with a medium in the various phases of its history. The realisation of living occurs in the interactions of the organism and the medium in a spontaneous flow of structural changes in which the organism and the medium change together congruently as long as the organism conserves its organisation and its adaptation to the medium through those structural changes. This dynamics of structural congruence that takes place between organism and medium is called structural coupling. The organism dies when the structural coupling is lost, when the organisation and the adaptation of the organism is lost.

MATURANA: For me these concepts are elements of an answer to the following question: How is it possible that we, as closed, structure-determined systems, can interact in a harmonious way? As all systems are structure-determined, an external agent cannot determine what happens inside them. The change is triggered by the perturbing agent but determined by the structure of the perturbed system. Instructive interactions are impossible. Of course, an external impact may lead to the disintegration of a system by destroying its organisation. It is also possible for the systems – due to structural change – to lose contact. They may, however, continue interacting by preserving some form of cohesion and maintain their organisation. Here we are concerned with the last variant of interaction.

POERKSEN: What is the foundation of such an encounter, of such continual contact between systems?

MATURANA: There must be some structural congruence. To use another everyday example: If you want to enter a locked room without breaking the door open or destroying the lock, you will need the right kind of key to gain access to the new domain. I would say, therefore, that lock and key must have a congruent structure.

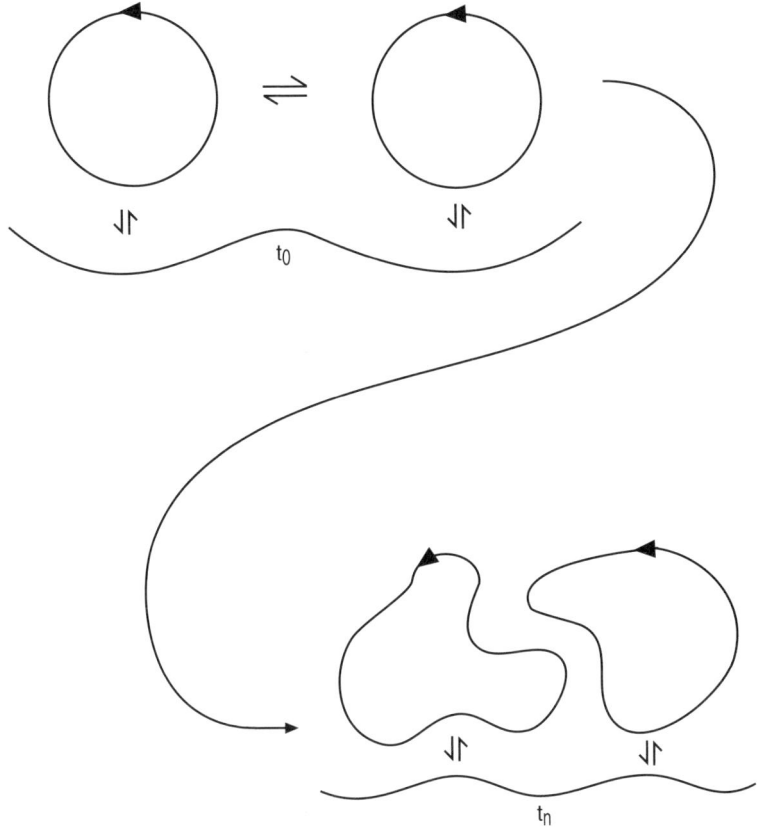

Fig. 8: The diagram shows two living systems and their interactions in a medium.

POERKSEN: Is this your answer to the question of how to enter a closed system? The motto would then be: Find the fitting key!

MATURANA: What matters is a specific relation between the lock and the key, which in this case is the result of planned production: someone designed lock and key in this particular way. – However, when a young man and a young woman find each other and, after a number of unimportant encounters with other people, fall head over heels in love, something very similar to the example of lock and key takes place. They look at each other – and stay together. Their particular congruent structures that enable them to enjoy their relationship is the result of our evolution that started billions of years ago.

THE MYTH OF SUCCESSFUL COMMUNICATION

POERKSEN: Why do you not want to use the current models of communication in order to explain, for example, the interview arrangements of two people? They have the advantage of being very simple and transparent: there is a sender, a receiver and a connecting channel. Communication and mutual orientation function through verbal and non-verbal sign or symbol systems that help to achieve the transmission of information.

MATURANA: Of course, we can describe how we pick up our telephone at a certain moment during the day, how we note appointments in our calendar, and how we finally put the receiver back. Obviously, I can describe these observable actions by means of the current models of communication based on the idea of information transmission and then state that we have just agreed on a meeting and that communication has taken place. Nevertheless, this characterisation only refers to appearances, to what is visible, and does not help us in any way to perceive the internal system operations and their relations to the domain of interactions.

POERKSEN: What then is, in your view, the meaning of *successful communication* or alleged *information transmission*?

MATURANA: The belief that communication has taken place is the comment of an observer who has perceived a flow of recurrent or recursive interactions and has, therefore, observed structurally coupled living beings. Observers talking about information transmission also register a mutually modified interaction. They have invented a concept that is expected to permit an explanation of reciprocal behaviour; this behaviour, however, is the result of structural coherences that they overlook. Before long, these observers are confronted by the problem of having to explain misunderstandings and diverging perceptions: such common phenomena cannot always be attributed to the malicious refusal of recipients to process the information received in the expected conventional way.

POERKSEN: Why are you so dissatisfied with these models and descriptions? Could they not be refined through a closer inspection and analysis of language, our prime instrument of mutual orientation? Language allows us to communicate and achieve the fine-tuning of relationships by means of words and sentences. The linguistic symbols are, therefore, the medium of consensualisation.

MATURANA: My view is completely different. The phenomenon of language emerges from a particular structural congruence that has evolved through a history of interaction. Just reflect on the precondition for the emergence of language: a coordination of behavioural coodinations. I maintain that symbols are of secondary, not primary importance. The ur-situation of language use is an everyday one. Imagine a person standing on the edge of a two-lane road and trying to stop a taxi. All the taxis passing by in the right direction are occupied. The person finally manages to stop a taxi driving by in the opposite direction and, having successfully contacted it, gestures to the taxi driver once more, drawing a circle in the air.

POERKSEN: And the taxi driver turns round …

MATURANA: Exactly. The gesture makes him change the lane and collect the client. What has happened here? Well, you will notice immediately what has happened when you observe, for instance, that the person decides to use another taxi that has suddenly appeared and stopped on the right side of the road, and when the other taxi driver shouts at her: "Why are you taking that taxi now that you have summoned me over?" There was only one eye contact, and there were two arm gestures, but they were understood as utterances. What happened was a coordination of the coordination of behaviours. From the first arm gesture and the moment of eye contact, taxi driver and potential client are in a reciprocal relationship and bound. The second arm gesture, the circle drawn in the air, coordinates their coordination. In brief: Whenever there is such a *coordination of coordinations of behaviour* in the flow of interactions, there is language. These are the processes, I maintain, that are necessary for us to be able to say that language is used in a situation.

89

THE WORLD ARISES IN LANGUAGE

POERKSEN: Your key example is from the domain of human interactions. Many other living beings communicate among themselves, and even with other species. Do they use language? Or are only human beings capable of developing language?

MATURANA: According to the present state of our knowledge, we have to accept that only human beings can live in language. When we ask ourselves whether there might be other living beings living in language, we are bound to do this by way of speaking, through living in language. Even when grappling with the problem of whether there exists an observer-independent reality out there somewhere, we need language for a discussion of this sort – and that is, in fact, the reason why such discussions and assertions of existence are totally nonsensical.

POERKSEN: How would you then describe, for example, the strange dances of the bees? There is undoubtedly mutual orientation there: The bees inform each other, we are told, which direction to take, what sort of flowers to avoid as unrewarding, where to expect plenty of nectar, etc.

MATURANA: Obviously the bees coordinate their behaviour – but the crucial question is whether they also coordinate the coordinations of behaviour, whether there is the phenomenon of recursion. Does one bee tell the other that it has, sadly, flown in the wrong direction? If this were, in fact, the case we would have to classify them as beings living in language.

POERKSEN: If I understand correctly, you seem to be concentrating particularly on the effects of an utterance in order to grasp what is essential to language. In our ordinary discourse about language, however, we do not focus on a series of mutually related behavioural coordinations but refer to a system of symbols, which is used for purposes of communication. We are concerned with the meaning of concepts (semantics), the structures of words and sentences (lexis and syntax), and the goal-directed, situation-bound use of these concepts, words, and sentences (pragmatics). I want to repeat my question: What is your specific view of language?

MATURANA: The crucial aspect is: there is a recursion in this coordination of coordinations of behaviours, a cyclical operation that is applied to the results of its previous application. Why is this factor so important for my understanding of language? My answer: Whenever we can observe a recursion there is something new; whenever there are cyclical operations of this kind there are new phenomena.

POERKSEN: Could you illustrate this particular effect of recursion with an example?

MATURANA: When you move your legs as if you were running, nobody watching you will say that you are running and moving away. People may perhaps think that you are trying to do some pantomime. However, when you are actually changing your location by moving your legs, then everybody will realise that you have started to walk and to run. This means: The phenomenon of running emerges exactly when the cyclical movement of your legs is linked to the linear shift of the surface that your feet happen to be touching at the moment. One movement builds on the previous movement, the simple repetition of leg movements is transformed into a recursion – and a new phenomenon emerges: you are running.

POERKSEN: What does this interest of yours in the notion of recursion contribute to the understanding of language?

MATURANA: I claim that whenever we encounter a recursive coordination of behaviour, that is, a flow in coordination of coordinations of behaviour, we see that something new arises, namely, language. As language arises, objects arise, e.g. the taxis. What is a taxi? What I say is that carrying and driving around passengers as a configuration of behaviour coordinated by the second coordination of behaviour (first recursion), becomes that configuration of behaviour that in a third coordination of behaviour (second recursion) appears "named" taxi. This means that objects arise as coordinations of coordinations of behaviour that obscure the behaviours that they coordinate (as taxi obscures carrying).

POERKSEN: What is the advantage of this new way of understanding language that you are proposing?

MATURANA: It reveals that language is not an instrument of information transmission and not a system of communication but a way and manner of living together in the flow of the coordination of coordinations of behaviour, which does not contradict the structural determinism of interacting systems. Once this has been grasped, it becomes obvious that symbols are not the beginning of language but that, conversely, language is the origin of symbols. Everything is turned upside down. Let us return, for a moment, to our central example of the interview appointment that we discussed at the beginning of our conversation concerning the interaction of systems and the phenomenon of language. The telephone conversation we had before you came to Chile was no transmission of information from Hamburg to Santiago or from Santiago to Hamburg; the decisive result of that interaction was and is that two structure-determined systems – Bernhard and Humberto – achieved the recursive coordination of their behaviour, the coordination of coordinations of behaviour. And here we are now, sitting together.

6. Autopoiesis is living

CONFRONTATION WITH DEATH

POERKSEN: In the year 1944 the physicist Erwin Schrödinger published a small book that has since become a classic in the history of science. It is entitled: *What is Life?* Your own thinking has been intensely concerned with this question. You have developed a description of the living – as a biologist –, the theory of autopoiesis, which is still causing excitement in the scientific world. However, let us start at the beginning. Why have you been so deeply fascinated and obsessed by the question of what it is to be living? Was there a particular incident, some key intellectual experience?

MATURANA: In fact, there have been various incidents and different key experiences that have inspired me. You must realise that I was often very ill as a child; death was a constant companion in the days of my childhood. I fell ill with tuberculosis several times, and the threat of this disease made me think about the relation between life and death quite early. I remember writing a poem at the age of 14 years, which deals with the difference between a corpse and a rock, the corpse being different from the rock because it had been alive. The fact of being alive was, therefore, not a property of matter – but what was it then, I asked, if one can lose it?

POERKSEN: You are describing a dialectical pattern: In the encounter with our own death, we become aware of our craving for life.

MATURANA: You could say so. In the year 1949 I was in a sanatorium in the mountains, being ill with tuberculosis again, and I had strict orders not to exert myself in any way; the prescribed therapy was, in fact, not to do anything at all. In secret, however, I read two books. In Nietzsche's *Thus Spake Zarathustra* I discovered that wonderful story of the metamorphosis of the spirit, the spirit being transformed first into a camel, then into a lion, and finally into a child. The child is described as the first movement: If I ever got out of the sanatorium alive, I thought, I would be like a child, starting from scratch, at the beginning again. Towards the end of Julian Huxley's book *Evolution: The Modern Synthesis,* I came upon a chapter in which I read that evolutionary progress meant increasing independence of a living being from its environment. Human beings, therefore, appear to be the most independent and most advanced living beings. So there I was lying in my bed, completely dependent on my medium, unable to leave the sanatorium, ill, close to death, and knew clearly that Julian Huxley could not have been right.

POERKSEN: If I understand correctly, the confrontation with death led you to ask the question of the nature of life. Moreover, Nietzsche and Huxley offered answers, which you could relate to your own situation.

MATURANA: Indeed. Life, I said to myself, has no meaning, no sense, and does not follow any programme of evolutionary progress. My conclusion had a tautological ring: the sense and the purpose of a living being is just to be what it is. The purpose of a dog is being a dog, the purpose of a human being to be a human being. Anything affecting a living being and happening to it, it became clear to me, had to do only with itself. When a dog bites me because I have stepped on its tail, then it bites me because it wants to avoid pain. This means that living beings are autonomous, that they have defined limits, that there is a boundary marking what belongs to them and what does not.

POERKSEN: It has become customary in biology to answer the question of what being alive means by drawing up a list of necessary properties. To be alive means, it is said for example, to be capable of reproducing and to be able to move around. Why did you not find any such list satisfactory?

MATURANA: Because such a procedure does not tell when the list of features or criteria that are necessary to claim that a system is a living system is complete. We cannot know when such a list has been completed unless we already know beforehand the list of features that characterise a system as a living system. In the year 1960 a student asked me during a lecture what actually began four thousand million years ago so that we can claim now that living systems began then. The question embarrassed me considerably because I could not answer it. Therefore, I asked the student to come back a year later; I would then be in a position to answer his question. I kept asking myself, though, as I continued to ponder the question, how I might be able to decide to have actually found the proper answer. How could I be sure to have defined life appropriately by listing features like reproduction or locomotion, special chemical composition, or a combination of such features?

POERKSEN: The problem is how we might be able to prove that we have found all the central features.

MATURANA: Drawing up a list of features presupposes, strictly speaking, knowledge of all the potential features. Only those who believe that they already have the answer, although they are still looking for it, could possibly know when their list is complete. I was, however, searching for an understanding of living systems which did not require the enumeration and classification of all the components and processes involved. I was looking for a form of organisation common to all living systems, which had to be independent from their particular components and their particular structures.

POERKSEN: How did you come to develop the theory which has become widely known under the catchword *autopoiesis*?

MATURANA: My own thinking went through various stages. At first I spoke about systems without an external purpose, whose activities make sense only within their own being. These *self-referential systems* were distinguished from *allo-referential systems*, whose essential feature was that they served a purpose external to them. (A car, for example, is an allo-referential system: its sense and purpose is to serve as a vehicle of locomotion from one place to another.) The concept of reference did not really appeal to me, though, because it always involves a relation between different elements – and I did not want to describe a pattern of relations; I wanted to understand the processes of a system through the system itself. Therefore, I looked for a concept that would highlight the processes that ultimately resulted in the phenomenon of self-reference.

POERKSEN: You wanted your theory of the living to be alive itself.

MATURANA: I was both obsessed and fascinated by a characterisation of the living that could not be separated from the actual realisation of the living. Although I had read Erwin Schrödinger's book, my question was not what life *was*, but what essentially constituted a living system. I wanted to discover the configuration of processes, the specific molecular dynamics, which produced, as a result, a living system, a cell, for instance. What must happen for such a system to arise? Conceptually, at least, I wanted to create a living system; that was my goal.

POERKSEN: You wanted to play God.

MATURANA (laughing): I did not want to *play* God, I wanted to *be* God.

A FACTORY THAT PRODUCES ITSELF

POERKSEN: What happened next in the step-by-step development of your new theory of the living?

MATURANA: In 1963, in the laboratory of a microbiologist friend with whom I regularly discussed the developments in molecular biology, I had the decisive inspiration. The dogma in molecular biology was, at the time, that information travels from the cell nucleus to the cell cytoplasm. We asked ourselves whether it might not also travel the other way, from the cytoplasm to the nucleus; nobody had yet heard of retroviruses, so our question was quite justified. We designed experiments that were never implemented, but one of those days I drew a diagram on the blackboard and said to my friend: "The DNA participates in the synthesis of the proteins, and the proteins participate as enzymes in the synthesis of the DNA." My diagram had the form of a circle. When I looked at what I had just drawn on the blackboard I exclaimed: "My goodness, Guillermo, that is it! This circularity of the processes reveals the dynamics that makes living systems autonomous, bounded, and independent entities." I had found the conceptual basis of the phenomenon that was later termed *autopoiesis*. From then on, I described living systems as circular systems.

POERKSEN: We have now reached the last phase of this short scientific and historical prelude. How did the concept of autopoiesis finally come to be invented?

MATURANA: It must have been in the year 1970; I had met with my friend José Maria Bulnes who had written a doctoral thesis on *Don Quixote*. In this thesis he dealt with the dilemma confronting Don Quixote: to choose the path of poiesis (production, creation), or to involve himself in praxis (actual work), without paying much attention to the consequences of his actions. He finally goes for praxis and decides to become a wandering knight, and therefore decides against poiesis and writing novels about wandering knights. During that conversation it hit me: "That is the word I have been looking for: *autopoiesis.*" It means *self-creation* and consists of the Greek words *autos* (self) and *poiein* (produce, create). I had successfully condensed into a concept my idea of what essentially characterises a living system. There was the additional advantage that the term was completely unknown – in contradistinction to the somewhat cumbersome expression *circular systems* –, and that it focussed the attention on the result of the constitutive processes of systems that produce themselves as unities through their own operations. The product of the autopoietic organisation of a system is that very system itself.

POERKSEN: Can the concept of autopoiesis be specified in greater detail?

MATURANA: Living systems produce themselves within their closed dynamics. They share the autopoietic organisation in the molecular domain. When we examine a living system, we find a network producing molecules that interact with each other in such a way as to produce molecules that, in turn, produce the network producing molecules, and determine its boundary. Such a network I call autopoietic. If we, therefore, encounter such a network in the molecular domain, whose operations effect its own production, then we are dealing with an autopoietic network and, consequently, with a living system. It produces itself. This system is open to the input of matter but closed with regard to the dynamics of the relations that generate it.

POERKSEN: Perhaps an example demonstrating the autopoiesis of the living would be helpful at this stage. You have often referred to the cell as an autopoietic system. Would that be a compelling model?

MATURANA: In my terminology the cell is described as a molecular autopoietic system of the first order; consequently, a multicellular entity is an autopoietic system of the second order. The special thing about cellular metabolism is that it produces components, which are in their entirety integrated into the network of transformations that produced them. The production of components is, therefore, the condition of the possibility of an edge, of a boundary, of the membrane of a cell. This membrane, in turn, participates in the ongoing processes of transformation, it participates in the autopoietic dynamics of the cell: it is the condition of the possibility of the operation of a network of transformations that produces the network as an integral whole. Without the boundary of the cell membrane everything would dissolve into some sort of molecular slime, and the molecules would diffuse in all directions. There would no longer be an independent entity.

POERKSEN: This means that the cell produces the membrane and the membrane the cell. The producer, the act of production, and the product, have become indistinguishable.

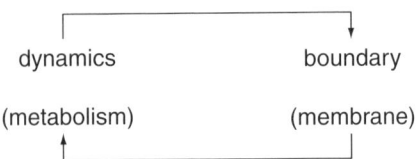

Fig. 9: The cell – an autopoietic system of the first order – is a factory that is its own product.

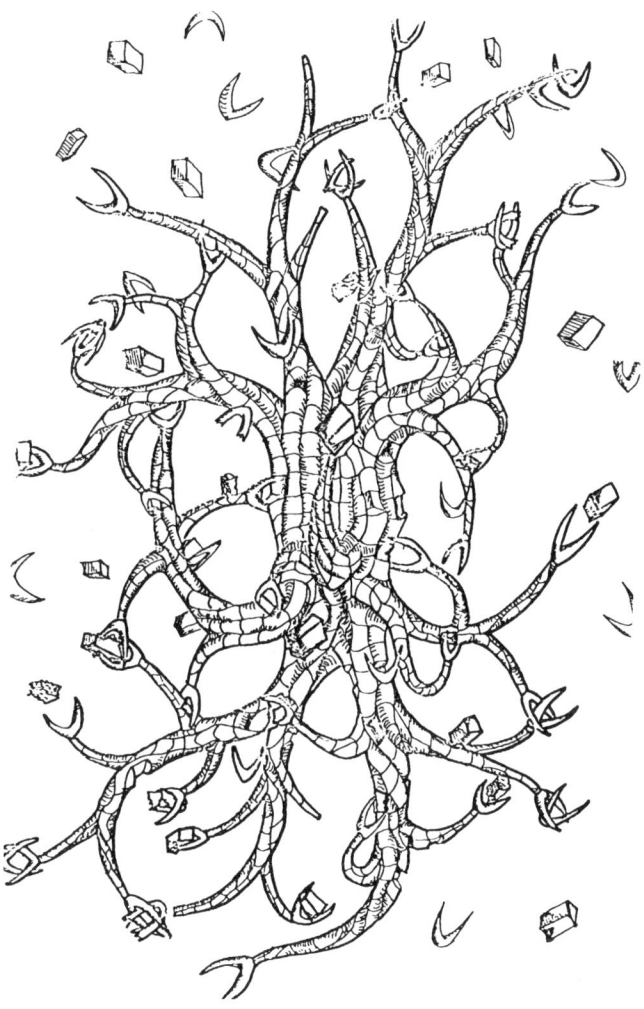

Fig. 10: An autopoietic system uses its components as elements of self-creation. (Drawn by Alejandro M. Maturana.)

MATURANA: I would say, a little more rigorously: The molecules of the cell membrane participate in the realisation of the autopoietic processes of the cell and in the production of other molecules within the autopoietic network of the cell; and autopoiesis generates the molecules of the membrane. They produce each other, and they participate in the constitution of the whole.

Autopoietic and allopoietic systems

POERKSEN: The specification of autopoietic systems of the first order I find convincing but I do not understand how you can say that autopoietic systems of the second order (e.g. human beings) can produce themselves. Could we not say just as well: Human beings, in their everyday lives, essentially produce what is different from them. People work, build houses, bake bread, knit sweaters, etc.

MATURANA: Naturally you can see human beings in this way. When you describe persons in the social domain as workers, then it is certainly possible to consider them as producers of bread or sweaters and characterise them primarily in this way. That they are living systems is practically irrelevant in this context because machines that make the same products could replace them.

POERKSEN: To classify an entity as an autopoietic system of the second order might, therefore, be the result of the approach selected and the perspective chosen?

MATURANA: Yes and no. The microphone that we use to record our conversation cannot simply be regarded as an autopoietic system even if we were very much tempted to do so. Only children can do that sometimes. In their play, the non-living may appear to be living. That is play, however, and we know that it is.

POERKSEN: The possibility of changing perspectives is only successful with one direction, then: We may view autopoietic systems as systems that produce something different from themselves. In the case of non-living systems such an approach does not work, they cannot be classified as autopoietic (simply because we feel like doing so).

MATURANA: Right. If I describe our microphone as a living system then you will certainly want to know how the autopoiesis of this system functions. And I shall not be able to give you a satisfactory explanation.

POERKSEN: What do you call systems that create something different from themselves?

MATURANA: Originally I called them, as already mentioned, allo-referential systems, today I call systems of which we can say that the sense and purpose of their operations are external to them, *allopoietic systems:* the result of their operations is not themselves – just think of cars and computers. This is, however, not at all a depreciative term, and it should not be misconstrued as the expression of a discriminatory hierarchisation. Without my car and my computer, I could not live the life that I would like to be leading.

POERKSEN: Is the feature of autonomy central for the realisation of autopoiesis? We could certainly claim that practically all systems are autonomous because they all function according to their own rules. If I shout at the waiter in a café at the top of my voice that I want to order a coffee, he may still fail to bring me one. The same will happen if I ask my coffee maker (an allopoietic system) quite politely to make me a cup of coffee. Coffee will be produced only when a filter is put in, when water is filled in, when the machine is switched on, – in brief, if I play according to the rules of the machine.

MATURANA: Naturally there are various possibilities for different systems to be autonomous, to function according to their own rules. Of course, there are many autonomous systems which are not living systems. It would be mistaken to see autonomy as the key feature of autopoiesis. The central point is that we have a closed network producing molecules that, in turn, produce the network that produced them. Put in a formula: Autopoiesis is the specific way and manner intrinsic to living systems of being autonomous, of realising their autonomy. Autonomy is the more general notion.

POERKSEN: How do we know that autopoiesis, this particular form of circular organisation, really is the crucial criterion of life? How could you prove that?

MATURANA: We could prove it by successfully presenting a series of processes that would as a result produce what somebody wants to have proved. We would have to demonstrate that the realisation of autopoiesis is the direct or indirect source of all the characteristics of living systems and ultimately produces an entity that has all the known and unknown features of a living system.

THE SECOND CREATION

POERKSEN: You designed a computer model that simulates an autopoietic system. In the scientific literature there has been the odd voice claiming that this simulation actually refuted your theory. Your simulation, the argument runs, is evidently not alive, although it may possess the features of a living system.

MATURANA: I can only reply that that model was meant to be an illustration and does not claim to prove anything. It certainly is no living system. The computer functions like the puppeteer in a puppet theatre, it is used to transform the different elements into entities that exhibit in the space of observation, i.e. the graphic space, a dynamics comparable with the molecular dynamics. The computer and its program are used to drive the elements which move autonomously in a living system: molecules need no puppeteer, they need no hidden power to move them, they move themselves – by thermal agitation due to thermodynamic laws. That is precisely what is special about them. However, as you know, immense efforts are invested at present to create artificial life. One day these attempts will undoubtedly succeed, although they hold terrible risks, and we will be able to build autopoietic systems in the molecular domain.

POERKSEN: If your prediction is borne out and artificial life is created, God would not simply be dead, as Nietzsche once said, God would have become superfluous, eliminated by the creation of autopoietic systems. Would you agree?

MATURANA: Not at all. Before we can answer such a question or formulate a thesis of this kind, we must have agreed on the meaning of the word *God*. Yogananda, the great Yogi who came to America, once said: When you think that God is far away, then God is far away, when you think that God is very near, then God is very near. The word God designates a human concept that has gained enormous significance and power in our world. To many people, however, God does not appear to be, as the Christian doctrine prescribes, an intelligent and creative Being in whose image we were made. What is decisive for them is that speaking of God permits them to speak of an inaccessible presence and of a connection with the source of their existence about which we cannot actually speak at all. If I, therefore, see God as the source of everything then he can certainly never be superfluous. The fact that life forms itself under specific conditions is then an expression of God's existence.

POERKSEN: An author managed to let emotions run high in Germany for some time by concluding all his interviews with the same question, – a good question, I think: Does God exist?

MATURANA: I was asked myself at the end of a lecture: "Do you believe in God?" My answer was: "I exist in God's Kingdom." The questioner repeated: "Do you believe in God?" I answered again: "I exist in God's Kingdom." And he repeated the question once more: "Once again: Do you believe or do you not believe in God?" – "Would you like me more or less," I said to him finally, "if I answered Yes or if I answered No?" His insistence was due to a desire for discrimination.

POERKSEN: And your answer meant, in fact: God's existence is not a matter of faith.

MATURANA: I would say that people who believe in God are plagued by severe doubts.

7. The history of an idea

A CONCEPT BECOMES FASHIONABLE

POERKSEN: You have been trying hard to retain the concept of autopoiesis exclusively for the characterisation of the living. Nevertheless, your ideas are now commonly used in social theory, in the description of society. Meanwhile, everything is an autopoietic system – science, journalism, football, families, art, politics, societies, etc. –, everything vibrates along according to its own rules within its own boundaries.

MATURANA: That is so. People like and honour me as the inventor of the term and the concept of autopoiesis – particularly so, when I am not present and unable to tell them what I really said. When I appear in person, however, I always point out that the concept is, in my opinion, only valid for a certain defined domain for which it solves a particular problem. A few years ago, for instance, I was invited to a conference at the *London School of Economics*, which dealt with the problem of whether social systems could be seen as autopoietic. The debate lasted three full days and, at the end, I was asked to say a few concluding words. I said: "For three days I have been listening to your ideas and exchanges, and I want to put the following question to you now: What are the features of a social system that would justify choosing as the topic of this conference the problem whether a social system could be classified as autopoietic or not?"

POERKSEN: You meant to suggest a different starting point for their deliberations: one must first understand the social phenomena before one can attempt to describe them more precisely with a concept borrowed from biology.

MATURANA: Precisely. Applying the concept of autopoiesis to explain social phenomena will cause them to vanish from your field of vision because your whole attention will be absorbed by the concept of autopoiesis. Naturally, we can discuss whether the house we are sitting in now is an autopoietic system. The choice of this topic, however, has the unavoidable effect that the features of an autopoietic system will guide our reflections. Asking for the constitutive properties of the entity of a house, however, and whether its characteristics accord with the concept of autopoiesis, will leave us free to analyse and investigate. We might then find that houses cannot be described as autopoietic, – or must be described as such. Who knows?

POERKSEN: Is it not fascinating to play with the idea of imagining a whole society as a collection of autopoietically functioning giant cells? One such giant cell, we might say, is formed by the media, another by politics; still others comprise the economic system, science, art, etc.

MATURANA: Naturally, in a community of artists works of art are produced, there are discourses and reflections about art, – but is all this autopoiesis? What is produced here in what domain and in what way? In all the different social systems you have quoted, there are indubitably dimensions of autonomy, but there is no autopoietic organisation. I can only repeat: Autopoiesis refers to a molecular network of the production of molecules that through their interactions produce that very network and create its boundary. Autopoiesis is *one* variant of autonomy among many others. Both concepts have to be strictly distinguished.

IMPLORING ERICH JANTSCH ON BENDED KNEES

POERKSEN: Your readers and disciples do not share your plea for precision. The astrophysicist Erich Jantsch, in his book *The self-organisation of the universe*, published in the late seventies, describes practically any kind of recursive figure as autopoietic. I am told that you met Jantsch once and, in a dramatic gesture, fell on your knees and implored him to stop misusing your concept. Is this correct?

MATURANA: It is. On that occasion, I tried to support my argument with a joke and to plead for a little more seriousness in a funny way.

My genuflection took place in the year 1978; Francisco Varela had organised a meeting with Heinz von Foerster, Gregory Bateson, Ernst von Glasersfeld, Erich Jantsch and me. We had dinner together – and at some stage, I knelt down and said to Erich Jantsch that he would destroy the idea of autopoiesis if he went on using the concept in such generality.

POERKSEN: How did he react?

MATURANA: He insisted that autopoiesis was very well suited to describe all systems that were autonomous in some way; my objections were not valid; I was not prepared to accept all the consequences of my own theory. My own view is, however, that using a concept outside its proper context of application means committing a double fault: the concept will work properly neither in the original nor the new domain.

POERKSEN: In Germany, the sociologist Niklas Luhmann at Bielefeld University has been one of the best-known proponents of the theory of autopoiesis. He introduced the concept in his central work *Soziale Systeme*, published in 1984, and from there went on to elaborate this theory by describing all the different domains of society as self-directed producers of their own specific realities. Luhmann brought about the *autopoietic turn* in sociology.

MATURANA: When I was a visiting professor at Bielefeld I never withheld my criticism but articulated it frequently in numerous debates. "Thank you for having made me famous in Germany," I said to Niklas Luhmann, "but I disagree with the way in which you are using my ideas. I suggest that we start with the question of the characteristics of social phenomena. The concept of society historically precedes the idea of the autopoiesis of living systems. Society was the primary subject of debate; autopoiesis and social systems came much later. It follows, therefore, that we should first deal with all the relevant phenomena appearing in the analyses of society and only afterwards ask ourselves whether they may be elucidated more precisely in terms of the concept of autopoiesis."

POERKSEN: You are cautioning against the dangers of reductionism.

MATURANA: The problem simply is that Niklas Luhmann uses the concept of autopoiesis as a principle in the explanation of social phenomena, which does not illuminate the processes to be described nor the social phenomena but tends to obscure them. Autopoiesis as a biological phenomenon involves a network of molecules that produces molecules. Molecules produce molecules, form themselves into other molecules, and may be divided into molecules. Niklas Luhmann, however, does not proceed from molecules producing molecules; for him everything revolves around communications producing communications. He believes that the phenomena are similar and that the situations are comparable. That is incorrect because molecules produce molecules without extraneous help, without support. This means: Autopoiesis takes place in a domain in which the interactions of the elements constituting it bring forth elements of the same kind; that is crucial. Communications, however, presuppose human beings that communicate. Communications can only produce communications with the help of human beings. The decision to replace molecules by communications places communications at the centre and excludes the human beings actually communicating. The human beings are excluded and even considered irrelevant; they only serve as the background and the basis into which the social system – conceived of as an autopoietic network of communications – is embedded.

POERKSEN: What swims into focus if we follow this perspective and describe a social system as a network of autopoietically self-reproducing communications is an extremely weird social structure: a society without human beings.

MATURANA: That is precisely the form of description manufactured by Niklas Luhmann. His conception can be compared with a statistical view of social systems: people with particular features do not feature in it. When we speak about social systems in our everyday life, however, we naturally have in mind all the individuals with their peculiar properties, who would protest against their characterisation as autopoietic networks – and do so, anyway, when they criticise Niklas Luhmann.

POERKSEN: One might of course say: Well, that is the objection of an empiricist, which need not upset a theoretical sociologist.

MATURANA: But all those who do not want to float about in a sphere of abstractions will surely insist on an answer to the question, How do we know that we are really dealing with a social system? Is it a social system because we are observing communications? Eventually human beings will inevitably emerge in the search for an answer. – But why does Niklas Luhmann proceed in this way, at all? He once told me that he excluded people from his theoretical design because he wanted to formulate universal statements. If one speaks about human beings, his argument was, universal statements become impossible. That view I do not share either.

HUMAN BEINGS ARE INDISPENSABLE

POERKSEN: The systems theory designed by Niklas Luhmann could perhaps be considered as a sort of *negative anthropology*: We cannot but remain silent in gentle humility and reverence regarding the infinitely manifold and ineffable mystery of humanity, the object of worship.

MATURANA: I do not believe so, because Niklas Luhmann has chosen this form of description to make universal statements. That was his reason for choosing a decidedly formal mode of description – like a mathematical system. What happens when human beings turn up with their likes and dislikes, their predilections and disinclinations, their desires and emotions? They are a threat to the beauty of the formal description and endanger the elegance of the formalism.

POERKSEN: Nevertheless, the refusal to convert human beings into elements of one's theory could also be interpreted as a particular form of appreciation.

MATURANA: That is possible; but even in the face of such a proposal you will have to take account of the people who may possibly complain and protest against their characterisation. If you deprive people of this opportunity, you treat them like freely disposable objects; they have the status of slaves, compelled to function without the opportunity of complaining when they do not like what is happening to them. Such treatment and contempt of people is standard practice in certain companies, communities and countries that negate individuals. A social system that forbids and even principally excludes complaint and protest is not a social system. It is a system of tyranny.

POERKSEN: If I understand your criticism correctly then it is primarily motivated by ethical considerations. This means that we are leaving critical epistemological analysis and entering the area of ethics. The issues are the protection of the individual and the fight for the rights of the individual.

MATURANA: Just imagine for a moment a social system that is, in actual fact, functioning autopoietically. It would be an autopoietic system of the third order, itself composed of autopoietic systems of the second order. This would entail that every single process taking place within this system would necessarily be subservient to the maintenance of the autopoiesis of the whole. Consequently, the individuals with all their peculiarities and diverse forms of self-presentation would vanish. They would have to subordinate themselves to the maintenance of autopoiesis. Their fate is of no further relevance. They must conform in order to preserve the identity of the system. This kind of negation of the individual is among the characteristics of totalitarian systems. Stalin, therefore, forced party members who did not share his outlook to give up their positions so as not to endanger the cohesion and the unity of the party. In a democratic form of communal life, however, individuals are of central relevance and, in fact, indispensable. Their properties create the unique character of a social system.

SYSTEMS THEORY AS WORLDVIEW

POERKSEN: The concept of autopoiesis has created a furore not only in science and amongst the followers of Erich Jantsch or Niklas Luhmann but also gained huge popularity in the New-Age scene. I think we are witnessing a sort of paradigm change with the theorists and opinion leaders of the New Age. Years ago they were attracted by modern physics and the dance of the atoms. It used to be reported that the physicist Werner Heisenberg, the creator of the Uncertainty Principle, and the Buddha practically shared an identical view of the essence of matter. The syncretism that emerged could be called *quantum theology*. For some time now, the key concepts of the New-Age scene have been provided by Gregory Bateson, Francisco Varela and – Humberto Maturana. The protagonists of the scene – Capra & Co. – have been brewing a rather explosive mixture of spiritualism and science, a sort of *network theology*, which is supposed to be the scientifically legitimated foundation of the worship of universal connectedness.

MATURANA: We have now hit upon the problem of reductionism, which is characteristic of our culture. Just look out the window for a moment. Over there, you see a loving couple, a young woman and a young man kissing each other. What is happening there? My answer would be: Whatever happens there happens in the domain of human relations. Naturally you can point out that in such exchanges of tenderness hormones and neurotransmitters are involved; no doubt we can speak of systemic processes in both organisms. All that would be correct, but what is occurring in the encounter of those two people, their feeling of love is not grasped or described by the reference to such processes: the loving tender relation that the two of them are living cannot be reduced to hormones, neurotransmitters and systemic processes. What they live occurs in them in the flow of their interactions as these give rise to the flow of what they do with each other through them. – When Fritjof Capra and others promote their quantum theology or some network theology and begin to worship systems or networks, they are thinking and arguing in a reductionist way. They flatten and blur everything. They no longer speak of molecules but only of systems that they elevate to their new gods. This is obviously reductionism, too. What I do is fundamentally different from a reductionist approach. Since I am always aware of the existence of different non-intersecting phenomenal domains, I take care not to confuse them in my thinking or in my writing. Indeed, if one does this, one can see that the phenomena of one domain cannot be expressed in terms of the phenomena of another domain. Thus, what happens in the domain of the operation of the organism as a totality in its relational space cannot be expressed in terms of the molecules that compose it, or vice versa. All that an observer can do is to see what happens in those two domains and attempt to establish a generative relation between them. I preserve, and attend to, the differences between the separate phenomenal domains in my descriptions. In this way, one sees the domain of molecules, the systemic domain, the domain of relations, etc. All these different domains constitute their own specific phenomena.

POERKSEN: Although I am not particularly inclined to defend the New-Age scene against anything, I think that it is no accident that your work has become attractive to that scene. The thesis of the observer-dependence of all knowledge can be interpreted as the removal of the subject-object rift that we encounter in the description of spiritual and mystical experiences.

MATURANA: These spiritual experiences have, in my opinion, nothing to do with experiences of transcendence in an ontological sense but much rather with an extension of awareness and an intensified feeling of participation: You become aware of being all at one with other human beings, with the cosmos, the biosphere, etc. When people talk about spiritual matters, however, they generally refer to some experience containing an ontological understanding or a true knowledge of nature. Such insights are, in my view, impossible in principle. Nothing that can be said is independent from us.

POERKSEN: Have you yourself had experiences that might be described as spiritual in your sense?

MATURANA: As I have already told you, I suffered from tuberculosis as a young man. After having spent seven months in bed, I went back to my school to find out whether I could still complete the school year in the regular way and so avoid having to repeat it. It was in December and I – having just got out of my sickbed – had to listen to a presentation prepared by my fellow pupils concerning the menace of tuberculosis. They described the terrible risks of this disease and the extremely limited opportunities for therapy available at the time. While I was listening to them, I felt myself slowly beginning to faint and decided to observe this process of fainting. When I regained consciousness, I was in the middle of the room and heard the voice of my teacher who said that I was looking very green and wanted to know what had happened.

POERKSEN: What had happened?

MATURANA: I shall tell you how I experienced the situation. When I prepared to observe the process of passing out, I lost all feeling for my body. I had no body any more but was still aware of being alive and gradually disappearing – like a wisp of smoke floating quietly and silently through a room – in a glorious blue cosmos. I felt like dissolving into that magnificent blue, fusing and becoming one with everything. Then suddenly everything was over. My head ached, I was sick; I heard the voice of my teacher and came round. What does this wonderful experience mean, I asked myself. Had I seen God? Was it a mystical experience? Or had I been on the way to death? – In the following weeks and months, I read the few books that existed at the time about near-death experiences and studied the medical and the mystical literature. It became obvious to me that I walked a very thin line with all the different interpretations. Reading the medical books and accepting their statements led me to believe that I had experienced what it is like to die and the effects that are caused by insufficient blood supply to the brain. If I believed the mystical literature, my experience involved an encounter with God and the unification with the totality of existence. At the time, I opted for the medical interpretation of what had happened to me as a near-death experience.

POERKSEN: Are these two interpretations so very different? Death could be a metaphor telling us of the gift of a new beginning: the old personality is dying.

MATURANA: It was, in any case, an experience that transformed my life. This transformation and the element of the extension of awareness restored to my experience a spiritual, a mystical dimension that was not so clear to me when I was young and thought I had to choose between the two interpretations. I lost all fear of death; I stopped clinging to things and unreasonably identifying myself with them because through the encounter with death I had experienced my connectedness with the whole. I became more reflective and less dogmatic. This is not intended to mean that I want to describe myself as an illuminated being above all earthly ties, not at all. That experience was so penetrating that it changed my life. Everything is transient, I realised, nothing but transition. We do not have to defend anything, we cannot hold on to anything.

II. Application of a theory

1. Psychotherapy

THE VIEW OF THE SYSTEMICIST

POERKSEN: The psychotherapists, in particular, have received your concepts and models of thought enthusiastically – and have truly celebrated you at their congresses for quite some time. Around the mid-eighties, there was practically no journal devoted to family therapy that did *not* quote you. Moreover, from time to time it seemed as if every therapist who was interested in systemics and constructivism would change into an epistemologically dyed-in-the-wool sage. This strong interest in your work shown by psychotherapy strikes me as rather strange, though, because the activity of the therapist is, if you are to be taken seriously, completely incalculable. You maintain that human beings cannot be controlled in a linear way and that there can be no instructive intervention – and this conception immediately destroys the healing ideal and the efficiency-based thinking of a whole generation of therapists.

MATURANA: I would say that it is not *the* therapy that becomes meaningless but a particular therapeutic conception, which rests on the idea of linear causality. All those who claim to know the universally valid procedure for freeing people from pain and suffering will no doubt feel irritated by my views. Nobody is capable of determining precisely what is going on in another person, and nobody is able to carry out instructive interventions on a structure-determined system – a human being – and to determine with exacting precision how that system will behave when confronted with a certain insight or experience.

POERKSEN: There can be no doubt that all therapists intend to cure their patients. And all those who are possessed by the desire to heal are ultimately – this is my thesis – dependent on some trivial concept of causality: they need raw mechanical thinking to prevent their work from becoming meaningless and a completely unpredictable activity.

MATURANA: Naturally all therapists want to help. But even though they may be convinced that they employ their techniques in an exact and goal-directed way, the desired effects will not necessarily arise. Whatever it is they do, it will only unfold its potentially healing effects outside the therapy room in a domain of human relations that differs from the world of images, conversations and experiences created in the room. The intentions and theories of human change that therapists may possess have no predictive quality because their considerations and wishes cannot be directly implemented in the domain of an individual client's human relations and lead to specific results. Therapists can merely classify the suffering of the people consulting them according to established categories and then convince themselves that a certain way of proceeding would be appropriate. However, this is not absolute knowledge.

POERKSEN: Could it be that your work is so popular in the therapeutic scene because it may be used as a theory of relief? A well-known psychotherapist writes with reference to your work: "Having abandoned the myth of instructive interaction, constructivist therapists may as well abandon the idea that they are responsible for the improvement and the healing of their clients." Conversely: Even if the client gets worse, the therapists are necessarily innocent. There is the perfect justification for therapists to extricate themselves from any critical situation.

MATURANA: We must argue more precisely with regard to this question. Of course, I cannot be held responsible for what others make of what I am saying or doing, for how they receive, understand, or interpret my actions and my utterances. They hear what they hear, understand what they understand, and do whatever they do. It cannot, in fact, be said that an utterance or an action has – due to a successful intervention – produced precisely the intended result in a person, for which one must now somehow carry the responsibility. In this respect I agree with the author you have quoted. This is, however, only one side of the coin: I cannot be made responsible for the actions of other people but I am definitely responsible for everything that I am saying or doing according to my own understanding, and for everything that I effect in the domain of human relations and in a systemic network. I may be acting with the intention of helping others or perhaps, alternatively, of cheating and manipulating them. These different intentions generate different kinds of action.

POERKSEN: The central demand you would formulate for the therapeutic community would then be: Abandon the idea of controlling and determining other people but declare yourselves responsible for whatever you do.

MATURANA: Certainly, yes. Those who have become aware that they cannot determine how other people will behave have also become aware that the quality of their actions depends on the extent of their wisdom. The wisdom of therapists manifests itself, I claim, in their ability to listen without prejudice, to display an attitude of openness and laissez-faire. Whatever seeks to express itself in a relationship is then not distorted by prejudices and personal leanings, by manipulative techniques or controlling desires, but perceived in the form in which it appears. To achieve this, one must listen with as many ears as possible, one must not allow one's perception to be blinded by premature judgments, and one must be aware of the emotions colouring what one is hearing. Whoever is curious, angry, envious or arrogant, will necessarily listen in a limited way, excluding further possibilities of encounter. Their attention is constricted by particular properties of the other person. The only emotion not limiting but enlarging one's listening is: love.

POERKSEN: Love is a dangerous concept to use with reference to the encounter between therapist and patient. Images of abuse immediately spring to mind; in any case, there is the fear of an improper loss of distance. However, this hasty judgment of mine is perhaps due to my refusal to listen to you at this moment.

MATURANA: If one attends to what people mean when talking about *fear, hate* or *love*, one can see that they always connote the particular domain of relational behaviour in which they do what they are doing at the moment, or in which they would like to do what they will do. That which we distinguish when we distinguish an emotion is a domain of relational behaviours. And as we distinguish different emotions we distinguish different classes of relational behaviours. Therefore, when we say that we are or that somebody else shows a particular emotion, what we mean is that we see him or her in a particular relational domain. Emotions constitute the relational fundament on which we do all that we do at any moment: all that we do we do in a particular relational domain, that is, in a particular emotion that gives to what we do its particular relational character as an action.

CHANGE OF CHANGE

POERKSEN: Are you suggesting that therapists should analyse their own feelings before beginning with their work?

MATURANA: What is necessary is not the analysis of the emotions, but the awareness of the relational dynamics that constitutes each particular emotion as an operation in a particular relational domain. It is in relation to this that I claim that the only emotion that does not constrain, filter or distort our vision or understanding, and that, on the contrary, expands our perceptive openness, freeing us from prejudices, ambitions or expectations, is what we call love in daily life. When is there love? Whenever we see a person behaving in a relational manner through which another being or him- or herself arises as a legitimate other in coexistence with him- or herself, we say that we see loving behaviour in that person. What is love, then? I claim that that which we connote in our daily life with the notion of love is the domain of those relational behaviours that we perform as a matter of course, through which another being that could be ourselves, arises as a legitimate other in coexistence with us. That is, in the relational domain of love the other is not asked and is not expected to justify his or her existence. He or she is seen, even in disagreement, without his or her presence being denied or obscured by prejudices or demands, so that his or her acceptance or negation is an act in seeing and not in blindness. Love is not an intended act of giving legitimacy to somebody else, love is unidirectional, and occurs as a spontaneous happening of accepting the legitimacy of the other as a matter of course without expecting retribution.

POERKSEN: Surely, this fundamental acceptance that you advocate cannot be practised without limits. In certain moments, it may be most effective for the therapists to enforce change by directed provocations and a measure of inconsiderate pressure.

MATURANA: Definitely. Acting on the basis of the emotion of love does not entail having to swallow any kind of behaviour or even consider it as personally relevant. The form of establishing relationships, however, is of crucial importance. The potentially frightening behaviour of the therapist will, if it is grounded in love, not be an expression of arrogance or of prejudice but a manifestation of profound and unbiased understanding. Patients may be shaken and shocked in order to free them of their blindness, and that is perfectly legitimate as long as it is done out of love.

POERKSEN: What consequences might your plea have for the requirement of therapeutic distance? Should the therapists who act out of love see themselves as members of the families that come to consult them?

MATURANA: The requirement is a double look: If therapists do not integrate themselves into a system to a certain extent they will not be able to listen properly, but at the same time they must keep a certain distance in order to remain in a position to see the context of what is happening and to maintain the freedom of reflection. A system can generally be specified as a network of relations. If people act within this network of relations, which constitutes the system, they have opted for a form of interaction that I call *agonal*: They act in a way that is in harmony with the established, traditional ways of behaviour of the system.

POERKSEN: What does this mean in particular?

MATURANA: When a mother complains to me about the awful behaviour of her daughter, for instance, and I then say to the girl that her behaviour is really quite awful and ask her for the reasons for such

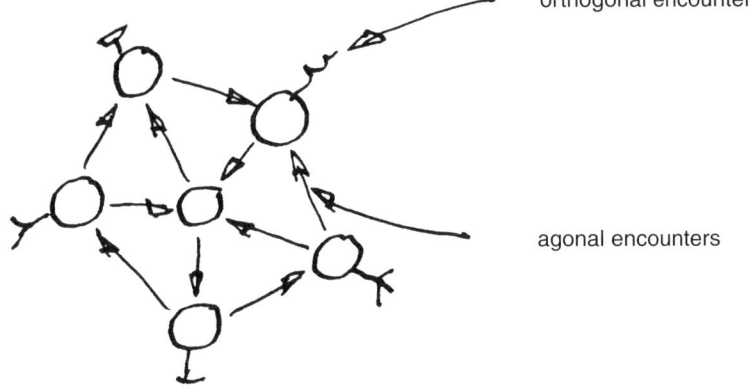

Fig. 11: A system (a composite unit) can undergo two kinds of encounters: (a) encounters in which an external entity meets the components of the system triggering in them structural changes of the same kind as those of the current manner of being of the system; and (b) encounters in which an external entity meets the components of the system triggering in them structural changes different from those appropriate to the manner of being of the system. I call the first kind of encounters agonal (confirming) encounters because the external agent acts in the encounter triggering in some components of the system the same structural changes that the other components of the system would trigger in them in the current dynamics of the system, so that the system remains in the same relational flow as a totality in which it was. And I call the second kind of encounter orthogonal (non-confirming) encounters because the external agent acts in the encounter triggering in some components of the system structural changes that are novel with respect to the current structural dynamics of the system, so that this undergoes a shift in the direction of its relational flow as a totality. (Drawing by Humberto R. Maturana)

behaviour, then I am participating directly in the interactions that have become entrenched in that system and therefore maintain it the way it is. An *orthogonal encounter* takes place, however, whenever people behave in ways that do not affirm the system but change its structure. The interaction is positioned, as it were, at a right angle to the dimensions participating in the creation and the maintenance of the system. The approach for an orthogonal encounter must be discovered through observation. Perhaps the mother complains with the expression: "This girl behaves awfully badly!" Then I begin asking questions about what really happened and finally start talking about the tremendous creativity of the girl. That is an orthogonal interaction that one must carry out – depending on the situation. Let me stress again, however, that the fundamental emotion of every therapist must be love. The path of healing is to discover anew self-respect and self-love.

POERKSEN: Could you use an example to demonstrate how such a therapy conducted out of love would function?

MATURANA: Well, as you know, I am not a therapist, so I cannot report about my practice but only about my everyday experience. Let me therefore choose a suitable example from that domain. – One winter day my little five-year-old grandson came to visit me. Due to his very poor vision, he has to wear thick glasses. That day he was also packed in many clothes to keep him warm. While playing in the garden, he slipped into the deep part of my pool. He went under but was pushed to the surface again because a lot of air had collected in his clothes. He desperately grabbed the pool edge and started to scream for help. I ran to the edge of the pool, pulled him out, and said to him: "Congratulations – you have saved yourself!"

POERKSEN: You re-interpreted the situation.

MATURANA: But not in an arbitrary way; that would not have worked. The boy had undoubtedly saved himself out of his own strength. Still profoundly frightened and terrified of being punished, he told me that it had all been an accident. "Of course it was an accident," I replied, "but you saved yourself, anyway. I just had to help you out of the pool." Sobbing, he mumbled that he urgently needed the toilet. "Oh, just wee," I said, "while I go and get a towel. The warm wee-wee will feel fantastic!" When his sister came to see me in the evening he ran up to her – and told her, beaming with joy and swollen with pride: "I fell in the pool, and I saved myself!" He did not feel guilty, he had developed no fear of the water, and he had not lost his self-confidence. You can consider this experience as a therapeutic intervention, if you like. There is the little boy with such poor vision that he falls into the pool by mistake but manages to save himself. And you do not react according to your own fear or anger but according to your perception and positive appreciation of the particular situation of the child.

Individual and society

POERKSEN: Perhaps it might be a good idea to try to apply your therapeutic attitude to the concept of resistance that is part of many therapeutic schools. What does it mean that a therapist diagnoses *resistance* in his clients? The client refuses, it is said, to get well, refuses to accept the mode of healing, and frustrates the positive effects of any therapeutic intervention.

MATURANA: If one defines as resistance the behaviour in a situation in which people decide to limit their own possibilities of perception so as to avoid seeing what other people want to show them, then one works with the attribution of guilt and with reprimands: The client's behaviour is judged negatively and condemned. Conversely, one might also say that those therapists who diagnose resistance have obviously not yet developed the form of listening, which enables patients to show their anxieties and reveal their epistemologies. However, if one is able to perceive that people show resistance simply because they are frightened, one may develop a different sort of understanding and feeling for the legitimacy of their behaviour: What this person does, one will then realise, is not directed at oneself.

POERKSEN: The world of psychotherapy is still ruled by one central distinction: the difference between individual and system and, in turn, by procedures directed at the individual or at the context. The name Maturana is unmistakably connected with systemic therapy, which requires not only the clients in need, but also parents, brothers and sisters, grandfathers or peers to be present in the therapy room. My question is now: Would you always prefer a system-oriented approach?

MATURANA: It seems to me that some systemic understanding will always be necessary because every action is embedded in a dynamics of relations. Although only the two of us are talking to each other here, we are not alone: our families, our cultures, our countries of origin, and our mother tongues are present in our conversations. Each of us carries with him a whole network of relations giving meaning to what and how we think, speak, or act. This is to say: Our encounter may be purely personal but we are both inextricably part of a systemic dynamics. Without an awareness of the formative power of culture, there is no mode of reflection that would enable us to establish what we actually do (according to our own decisions) and what just happens through us (due to our origins). Only the awareness of those shaping influences generates the opportunity for liberation.

Poerksen: You are painting the framework of these influences on a large canvas when you speak of the power of cultural conventions. I think this is very telling, because I have often asked myself why the systemic therapists, as a rule, limit their inspections to close relatives and do not extend them to the surrounding society, the constructs of a state or perhaps an entire world culture. And I have asked myself why this should be so because we are obviously not just shaped by our mothers and fathers, brothers and sisters. I could think of only one reason: you simply cannot send demands for payment to cultures.

Maturana (laughing): Quite possible, although one could, of course, make cultural influences visible, too, and then send bills to people who are prepared to pay for such work – so as to secure the therapists' livelihood. The pain becoming manifest in a therapy is always culturally conditioned, as has been clearly shown by the Chilean family consultant Ximena Dávila Yánez. It arises in a patriarchally oriented culture ruled by mistrust, possession claims, and the constant negation of other human beings. – People who are not heard and seen by their partners or at work, who have no presence where they live, experience this hidden rejection most painfully.

The Constructions of Pathology

Poerksen: After our discourses about the therapeutic effects of love and the power of culture, this may be a good moment for turning to another topic that is more closely related to your epistemological work. It is the critical question of what the concepts of mental health and of normality could possibly mean in the context of your ideas. Or the other way round: Psychiatrists say about patients who hallucinate that they have lost their "connection with reality", that is, they use an ontologically contaminated diagnostic formula because they base their diagnosis implicitly on a reality that can be known. What *you* are saying is that every reality construct is inescapably observer-dependent. What does it mean, therefore, to be ill or abnormal?

MATURANA: Let me give my answer in the following way: There are no pathologies in the biological domain. The cat is not an underdeveloped tiger; a tiger is not a pathologically arrogant cat. The tick sucking your blood is not in any way good or evil but simply lives along the way it lives – and unfortunately, you happen to be part of the leg it has sunk its proboscis in. This means that all forms of life have to be accepted as legitimate. To those who have chosen the path of *objectivity in parentheses,* pathology is not a feature of a world existing independently from an observer: an illness appears as a condition that observers – according to their inclinations – may consider undesirable. To be normal and healthy means, correspondingly, that one does not make any effort in the stream of life to change one's situation with external assistance. There is no *pathology as such,* there are no *problems as such,* and there is no illness that is independent of an observer's desires and predilections.

POERKSEN: Whose definition of normality and desirability counts or should count? We might, of course, imagine that certain persons enjoy their psychoses; their relatives feel obliged, however, to stop that sort of enjoyment by putting them away in a closed institution. The psychiatrist Thure von Uexküll once reported that the high fever accompanying a severe illness gave him some of the most beautiful experiences of his life.

MATURANA: There is no general answer to the question of what happens when someone who is labelled ill nevertheless feels very well. There is no stable criterion to direct our decisions here because every thing depends on the emotions guiding the actions. Perhaps people are scared by someone's supposed madness; perhaps people love an alleged patient and want to save him from certain death; perhaps some people talk in such a dangerous and revolutionary way that other people begin to fear for the loss of their privileges. People may hit upon the thought of declaring the demand for social equality to be a symptom of a special pathology against which society must urgently be protected. In the Soviet Union, as is well known, many dissidents were placed in psychiatric institutions on the basis of the simple argument that their ideas were pathological and that they could be delivered of their manic thoughts by means of electric shocks. The ascription of illness is an argument to stop all further debate.

Poerksen: What are the consequences of your views? Should all the psychiatric institutions be opened and should all the so-called patients be set free, since they are defined as ill only by certain observers?

Maturana: These people are plainly labelled pathological in order to shut them away. Let me be quite clear, however: I am not in favour of any sort of patient liberation; it might once more mean the implementation of a theory without proper knowledge of the specific situations. I would, however, see my goal in the achievement of an adequate awareness of the responsibility connected with the description of some condition as *ill* or as *not normal*: Such an attribution rests on a decision for which there is no higher reason, no absolutely valid legitimation, and no observer-independent justification.

Poerksen: You say yourself that you are no therapist but a biologist facing up to the fundamental questions of philosophy. It is a fact, however, that your reflections have been taken up most intensively in psychotherapy and related fields of person transformation – in pedagogy and management design. How would you explain this particular interest as well as your popularity?

Maturana: First I would like to note that this sudden rise in popularity did not affect me too much because I live in Chile and simply cannot accept all the invitations extended to me. It was most instructive, naturally, to become acquainted more closely with the work of the family therapists. The admiration of other people, however, has always made me query why and with what degree of understanding I am admired, and what will happen if people come to the conclusion one day that I do not actually hold the grand ideas that are so enthusiastically read into my papers. I assume that many therapists were fascinated by the fact that my work in biology made it possible to understand a family as a multiverse of different realities in which all the different member individuals can be right at the same time although their statements contradict each other.

Poerksen: The eighties were the era of theoretical debate in the psychotherapeutic scene; today the practitioners and the pragmatically inclined fine mechanics seem to dominate the scene again. You achieved instant fame through an explosively delivered lecture by American family therapist Paul Dell in 1981. At a symposium in Zurich he hurled his new creed at the audience with the zest of the newly converted: "There is no information", he was heard to proclaim. "An illness as such does not exist. Knowledge of truth is impossible." And so on. What is your relationship with the world of psychotherapy today?

Maturana: My popularity has in the meantime decreased noticeably, and that is perfectly understandable in a culture which is possessed by an insatiable appetite for new things, which presses for the practical implementation of everything with the intention of creating a method that will generate predictable results in the most efficient way. I am completely unsuited to promote this kind of efficiency-oriented thinking because my work demonstrates that such a universally functioning method of human engineering can never and under no circumstances be attained. We always encounter the other human being in a domain of fundamental uncertainty and all we can do is to try and create a form of existence, which allows us to dance together. One day – assuming we are therapists – the client will feel transformed and begin to manage his life again without external help.

Poerksen: A last question or note. As I see it, the application of your theories in therapy or in management design always takes place between two extremes. One extreme: Your followers act with a new awareness of complexity, guided by a feeling of modesty as well as the deep understanding that we cannot change the world according to our own conceptions and principles of control but must take into consideration all the relevant participants, the whole network of interacting elements that itself can never be penetrated completely. The other extreme: For some time now, your theories are increasingly traded as the instruments of ultimately successful manipulation. The motto is: *Because* we understand how closed systems can be perturbed, we can make successful use of this insight. I think that these two extremes of application may be found in the professional biographies of many systems theorists. More drastically: Some time ago we were moving close to a new kind of mysticism, today we are management consultants; some time ago we were searching for a new spiritual orientation, today we prefer making good money.

MATURANA: If your assessment is correct, then I can only state very clearly that I see a distortion of my thinking. In those attempts at manipulation, my work is not used to create a more humane form of life but exploited for personal advantages, for the enrichment of individuals, to serve the ruling idols of our culture, efficiency-orientation, control mania, and the craving for success. There is nothing else for me to do but accept these developments and keep up my trust in human nature, hoping that other people may use my work for the good of as many human beings as possible. If I tried to prevent the abuse of my ideas I would inevitably turn into a tyrant – and negate the biology of knowledge and the biology of love.

2. Education

THE PARADOX OF EDUCATION

POERKSEN: Immanuel Kant writes in his essay *Über Pädagogik* that the wide field of education is governed by a fundamental paradox. On the one hand, we want free and self-determined individuals to leave our schools, on the other we impose a syllabus on the future individuals, force them to attend schools, punish their failures, and persecute their non-compliance. There is, if we follow Kant, an inescapable relationship of tension between the goals and the means of educational efforts: They contradict each other. Would you agree?

MATURANA: No. Education, the commentary of an observer, is the process of transformation resulting from the co-existence with adults. We become the adults we have been living with. This means: If freedom and self-determined thinking are the goals of educational activity, then we have to live together in a way that is supported by the mutual respect for the autonomy of the other. In my view, the paradox formulated by Kant does not exist at all. The way of life, the manner of living together, shape and transform people. If you want to teach autonomy and reflection, you cannot use force as a method but must create an open space for communal reflection and action. There must be no contradiction between goals and means.

POERKSEN: Surely, there must also be constraints? It must be laid down when everybody has to be present, what the task is, who the teacher is, who has authority.

MATURANA: Compulsion will emerge if the teachers do not succeed in presenting their material in a thrilling way and in making school an attractive place of being together. Their failure will lead to force.

POERKSEN: The teacher is totally responsible for everything that may happen in school. Is this not an exaggerated claim?

MATURANA: No. If a teacher behaves respectfully, if he does not intimidate his pupils, if he listens, encourages cooperation and reflection, then a special form of interaction will emerge. The way of living practised by the teacher, including the goals of teaching, will be the source of profitable learning for the pupils. This also implies that three questions and tasks must be sorted out cooperatively in education. First it seems necessary to me to debate the educational ideal to be chosen – what should the future adults be like when leaving the school one day? Should they be democratically minded and responsibly acting citizens? Or are they to be authoritarian and commandeering hierarchs, *lords* who feel superior to everyone else? It is then necessary to anchor a way of life in the school that permits acting and thinking according to that ideal. Finally, there is the essential task of preparing the teachers for their job in such a way as to do justice to the desired goals – to enable them to live what they have to achieve.

POERKSEN: This would mean that teaching has nothing to do with the step-by-step elimination of ignorance, as is commonly thought. The transmission of knowledge is secondary. The primary requirement is a way of life that corresponds with one's ideals, a particular form of living together, out of which the material topics will arise in due course.

MATURANA: Exactly. The children do not learn mathematics in school, they learn how to live together with a mathematics teacher. Perhaps they will one day carry on this enjoyable and exciting kind of being together independently – and become maths teachers or mathematicians themselves. Teachers do not simply transmit some content; they acquaint their pupils with a way of living. In the process, the rules of arithmetic, the laws of physics, or the grammar of a language will be acquired. My claim is: *Pupils learn teachers.*

POERKSEN: What about children who systematically refuse to cooperate? What is to be done with them? The classical answer is, of course: bad marks, relegation, exclusion from the winning circles.

MATURANA: The so-called difficult children about whom teachers keep complaining often only struggle to be seen and accepted while the whole world expects them to behave in a calculable manner and to adapt to strange demands. Asking these children what they would like to do opens up space for an exchange and the children will give up their resistance. It is profoundly healing to be actually seen, to regain self-respect, and to participate in interactions supported by love. Some pupils probably turn away because the kind of teaching offered appears useless and boring to them. If the father is a bricklayer and the son is destined for the same trade – why should he waste his time on advanced algebra? Such an assessment on the part of the pupils is a challenge to the skills of the teacher who must now prove that the activity of a bricklayer can sensibly be connected with higher mathematics. – Everything is interesting once you are interested in it.

LISTENING TO THE LISTENING

POERKSEN: You think that anyone will become an enthusiast as long as the required talent for presentation is available.

MATURANA: Naturally. I remember vividly a teacher who came to one of my seminars one day. She found my ideas attractive, she told me, but she simply had to teach her children grammar – a laborious and dull kind of task. If she thought, I answered, that the teaching of grammar was inevitably laborious and boring and would not lead to a new understanding of one's language, then she would make her pupils feel precisely the same way. Of course, I could not tell her what she should do because she would have replied immediately: "That is what I have already tried many times – it just does not work!" However, I told her one thing very clearly: "If you have so little respect and love for your work, then your pupils are bound to hate grammar." She had to find the solution herself by changing her inner attitude; the pupils are always quick to grasp whether their teacher is enthusiastically involved in his task.

POERKSEN: But the problem surely is whether everyone will be interested in everything.

MATURANA: That is not the problem. Children are quite ready to become enthusiastically involved in anything provided of course that there are no people around who keep signalling and saying: "mathematics is tedious, grammar is dull, biology is uninteresting." People who have come to think in this way are handicapped. It is, of course, a permanent task to connect school topics with the everyday life of the pupils, and to foreground the questions that are of relevance to them.

POERKSEN: Are there not, however, certain amounts of subject matter that simply must be memorised because their deeper connection with the world of experience cannot be adequately dealt with owing to lack of time? The psychologist Ernst von Glasersfeld once suggested distinguishing between *training a*nd *teaching;* between drilling and learning by rote, on the one hand, and the active and creative construction of concepts and notions, on the other. Both are needed, he claims, and must be practised in the proper mixture as required by the given task.

MATURANA: A teacher must be flexible enough to choose the right procedure depending on the situation. Of course, certain things have to be memorised or practised repeatedly. Even simple repetition can, however, improve understanding because it sharpens one's vision and produces new insights. All of a sudden we find it easier to solve the equations before us; all at once our muscles change after we have dropped the ball into the net a few hundred times; our shots have become more precise. If you devalue the practice of repetition to an unavoidably boring routine activity, then you give it an additional significance that it does not deserve.

POERKSEN: Do you think that only good teachers make us really learn something? One morning there was a writing on the walls of the school I had attended: "We had bad teachers. It was a good school."

This seems to be quite correct to me because there is a dialectic of learning: You can definitely learn something by being exposed to negative examples, even including encounters with civil servants who are apparently immune to enthusiasm.

MATURANA: This is not the way I see it. The fact that some pupils obviously get along even under unacceptable circumstances does not suggest at all that the bad teachers are helping them in any way. Children who are confronted by disrespect and cruelty definitely need a space in which they can respect both themselves and others. A Peruvian psychologist has shown in a study that just one fully trusting adult is enough to help children regain their self-respect. Perhaps it will be the parents of the children who are maltreated by teachers, who believe in them, trust them and love them. Their support will help them to survive their terrible experiences and find their own way without despairing and breaking up. School cannot do too much harm in such cases. However, if there is no support from home and parents and no reinforcement of an autonomous way of life, then the schools have a special responsibility: Where if not there could a child develop self-confidence?

POERKSEN: You have recently established an institute in the centre of Santiago de Chile, which is mainly devoted to the further education of teachers. What recommendations do you offer the participants in your courses?

MATURANA: The distinction between two different kinds of listening seems to me to be of elementary importance for teaching. On the one hand, we can, if something is said to us, always check whether we agree with it, seek to establish the degree of correspondence with our own views – a central and wide-spread tendency in our culture. Those who listen in this way do not, however, really listen at all to anyone else except to themselves. The other kind of listening considers the question of the circumstances in which what is said is valid. In what area of reality is it correct? Do I like the world that is produced here? The advice I give the teachers coming to my courses is to show enormous patience, to listen to their pupils intensely, and to listen to their listening. If they respect others and grant them a space of legitimate presence, they will become loving beings in the flow of interactions. What do children actually hear, we must ask, when we talk to them? What do they perceive? Are they scared of an act of aggression? Do they feel confronted by a threat? Or do they feel invited to cooperate?

PERCEPTION AND ILLUSION

POERKSEN: In our present schools, mistakes appear to be of immense importance as indicators of failure and symbols of inadequacy. Schools, we might say, are training institutions for the avoidance of mistakes: they punish errors, mark wrong answers in red ink and reward faultless perfection with best marks. My question is now: What, in your view, are mistakes? How would you comment on this orientation of school practice?

MATURANA: We must see clearly that all human beings are intelligent and that they only very rarely commit logical mistakes. Children, in particular, make use of numerous distinctions, which do not please the adults and are, therefore, declared false and questionable. The opinion, for instance, that the ideas of a pupil are illogical and false means, as a rule, nothing but that what was said belongs to a domain of logic that is different from the logical domain which is the basis of the observer's listening and judging. In other words, a mistake is a statement made in a particular domain of reality, which is heard and evaluated in the context of another.

POERKSEN: It is commonly assumed that making a mistake means *not yet* knowing or seeing correctly.

MATURANA: Those who take the path of *objectivity without parentheses* consider mistakes and illusions as punishable errors, as symptoms of failure: There is something to be perceived and understood, and people simply cannot do it, cannot see things the way they really are. Those who, in contradistinction, follow the path of *objectivity in parentheses*, take the experience of an illusion and a mistake seriously. They want to know how illusions and mistakes arise. The answer is: Something is triggered in a structure-determined organism that corresponds to the proper features of the apparently perceived phenomenon in a certain way and from a restricted perspective. This means that we can see illusions and mistakes – ironically – as partial truths; they partially correspond to a phenomenon, but we believe operationally that they are identical with the entire phenomenon.

POERKSEN: Could you give us an example?

MATURANA: Just think of the trout jumping to catch the artificial fly of the angler. It does so because the feathered hook is a perfect imitation of an insect hovering above the water surface. The realisation that there is no fly arises later, when the trout is dangling from the hook. The experience of an illusion is, as the example shows, accepted as valid when it actually occurs; it is devalued only because of other experiences and then classified as a perceptual error. In brief: Illusions and mistakes arise after the event, *a posteriori*.

POERKSEN: Are there not perceptions that are patently illusory? What if I said to you: "Professor Maturana, look outside, there is a unicorn at the window observing us."

MATURANA: There are different possibilities of reacting to what you are telling me. I might suspect that you are making fun of me, or I might assume – unicorns are, to the best of our knowledge, mythological entities – that you are, at the moment, suffering from hallucinations. It is also possible that I might interpret your pointing out the unicorn there as an attempt to start a discussion about the indistinguishability of perception and illusion. All these interpretations have, however, one thing in common: they devalue the experience you have described to me.

POERKSEN: Could we just for a moment assume that I am really seeing a unicorn?

MATURANA: Of course we can. We should then discuss why I cannot participate in your experience – why I do not see the unicorn observing us. Is my perception limited in some way? Or is that unicorn possibly part of your internal world, which is inaccessible to me? However, I really want to point out something else: I claim that it is impossible, *in the moment of the experience*, to distinguish between perception and illusion. If you are seriously reporting to me that there is a unicorn outside the window, then you are totally living in that world. Your whole body lives in that experience. You are absorbed by that world. Only later will it be possible to identify the unicorn as some strange movement of the leaves caused by a couple of birds. This means that an illusion is an experience that remains valid until it is disqualified by other experiences.

POERKSEN: So we actually never know whether what we see and describe is something real.

MATURANA: In the moment of the experience such a distinction is impossible in principle. We are always dependent on the reference to other experiences that can, in turn, only be classified as perceptions or illusions if they are related to other experiences. And so on.

POERKSEN: Does this mean that we might exist in a world of illusions throughout our lives without being able to ascertain this with any degree of certainty?

MATURANA: Immanuel Kant could formulate a thesis like that in connection with the *ding an sich* that we cannot know although it exists. You need the ultimate reference in order to be able to say that everything is an illusion. I would not argue in this way.

POERKSEN: What I actually wanted to ask is whether we can, in a deep sense, ever make sure that what we assume is not illusory?

MATURANA: We can never know whether our perception today will not appear as an illusion tomorrow. It may, of course, remain valid throughout our whole life. It is possible, after all, that I might confess to you tomorrow that everything I said yesterday was false. How can you know that, by the end of the week, your trip to Chile will not appear to you to have been a mistake? And that, when you are listening to the recorded tapes again, you will not come to the conclusion that Humberto Maturana talks complete nonsense?

POERKSEN: I certainly do not expect that because I have prepared myself intensively for our time together. I read your books, bought my ticket, booked a hotel. The sudden loss of all this stability and the collapse of my previous beliefs would probably upset me very much, and for that simple reason alone I would not be prepared to consider my trip to Chile a mistake.

MATURANA: Nevertheless, we simply do not know whether you will not one day end up with such an assessment. The crucial aspect is, however, that we always hold the experiences that we are making to be valid. In this sense you are, of course, right: We need this stability in the flow of our lives, we operate out of implicit trust, and we usually do not commit mistakes because we live along within the coherences of our structural coupling. Therefore, mistakes are infrequent and do not indicate some failure with reference to an observer-independent reality; they are *a posteriori* evaluations and reflections of a human being living in language.

ALL HUMAN BEINGS ARE EQUALLY INTELLIGENT

POERKSEN: For most of your academic life you have been engaged in research and not so much in teaching. Let me ask you nevertheless: What has working with students meant to you? From time to time, there is some debate in the universities about whether the combination of research and teaching should be abolished: The students, it is said, simply cannot cope, and the top researchers, at least, should be exempted from teaching.

MATURANA: I do not consider that desirable at all. Teaching has always been extremely important to me because, when inspired by intelligent remarks from students, I could use my seminars as laboratories for testing possibilities of thinking. I have never been bored, because any question that comes up may be interesting and may lead to further reflections when you look at it more closely. I could never accept a devaluation of the students because I fundamentally believe that all human beings are equally intelligent, anyway.

POERKSEN: Is this correct? Surely, some are a little more equal than equal – and therefore a little cleverer than others.

MATURANA: No. Intelligence manifests itself in the possibility of varying one's behaviour in a changing world. Whenever we classify a living being as intelligent we want to assert that it is capable of adequately transforming its life. As beings living in language we need and possess such a gigantic plasticity of behaviour that we can say with full justification: This one fact that we exist in a domain of the coordination of coordinations of behaviour renders all of us living beings of equal intelligence. Naturally, there are different experiences and predilections, interests and capabilities, – that is certainly the case. I claim, however, that any human being can learn what any other human being has learned, if he or she only wants to.

POERKSEN: Now you sound as if any individual could become an Albert Einstein – an icon of superior intelligence.

MATURANA: Not everyone can become an Albert Einstein but everyone can, if they want to, learn what Albert Einstein learnt and taught. Naturally, they will not go the same way as Albert Einstein, and they will not invent the same concepts and theories because this would require the same circumstances of living plus identical experiences. In addition, any person who has chosen a form of life and a professional career inevitably constrains his other capabilities. If I want to become a star in bodybuilding, I have to concentrate on particular demands – and others will not even surface. This does not mean, however, that the bodybuilder lacks fundamental intelligence simply because of his or her decision for a certain kind of existence.

POERKSEN: How do you explain, then, that all these equally intelligent human beings are not at all equally successful? Most of the intelligence tests on the market are based on the assumption that the differential success in solving the problems presented is an indicator of intelligence.

MATURANA: What intelligence tests elicit and diagnose is the degree of inclusion in a culture. It is, as I claim, the emotions that determine whether and to what extent we are able to exploit our capabilities and our fundamental intelligence. The dominant emotion modulates intelligent behaviour in a decisive manner. Some individuals may be unable to follow because they are scared, and they will behave differently from individuals who are depressed or who are just bored because their interests lie somewhere else. Finally, an enormous breadth of variation in predilections and capabilities arises from the particular situation in which people grow up. Were they loved when they were young? Were they properly looked after? Was there enough food? Anyway, I insist: Intelligence is, for me, not some specific activity but the general capacity to move in a changing world flexibly and with internal plasticity.

POERKSEN: Still, there is the undisputed experience that people try and work extremely hard to understand things and nevertheless just do not ever manage.

MATURANA: If people make every effort and work hard this might indicate that they are actually bored. Why should we occupy ourselves with certain topics? Only to be able to demonstrate that we are intelligent? What other purposes might the knowledge serve that we have to acquire for the intelligence tests? Perhaps the people worried by questions of this sort would do well to move into other domains that are of real interest to them and in which they are active with pleasure and concentrated attention. But perhaps they are blocked by fear; perhaps children are afraid of the teachers' punishment and tormented by the fear of failure as soon as they arrive at their school. In such cases love, respect and trust will help.

III. History of a theory

1. Beginnings and inspirations

INSIGHTS OF A CHILD

POERKSEN: Your theory has a circular design, it is circular: The observer and the observed, the knower and the known form an inseparable unity. Regarding a circle we realise: it has no beginning and no end unless someone cuts it somewhere and creates a beginning. It may seem somewhat inappropriate, therefore, to inquire about the beginnings and starting conditions of circular thinking. The form of the question contradicts the format of the theory. Nevertheless: What inspired you, who were the people that influenced you? Where would you like to start and create a beginning?

MATURANA: No doubt my mother had a great influence in my emotional and intellectual growth. She taught me to accept the responsibility for my own understanding of the world and to have confidence in myself. I remember one day I was playing with my elder brother and our mother summoned us (I was eleven years old): "Children!" she said, "Nothing is good or bad in itself. Some behaviour may be appropriate or inappropriate, right or wrong, and it is your responsibility to decide which!" She added: "All right, now go and play again!"

POERKSEN: Why is this episode important for you?

MATURANA: If some behaviour cannot be classified as good or bad in itself, then we must – I realised – pay attention to the web of relations in which it is embedded and autonomously choose our mode of action. For me, a particular attitude reveals itself here. It involves trust in my brother and me and the belief in the autonomy and the freedom of every human being, which must be dealt with respectfully: Nothing has an unconditionally fixed validity – and therefore it is necessary to weigh up, to choose, and to decide.

POERKSEN: Chilean society is socially divided and split into rich and poor: The people existing in the run-down sheds, the *poblaciones*, on the outskirts of Santiago, and those residing in the magnificent town villas of Providencia, live in totally different worlds. How did you grow up? Did your family belong to the relatively small social upper class?

MATURANA: We were poor, although other people had to survive in significantly worse circumstances. I shall never forget the day I accompanied my mother, who was a social worker, at work. She went to see a woman who was ill in order to check her degree of neediness and her eligibility for free medical care. When we reached the place where this woman lived – it was nothing but a hole dug in the earth and covered by a roof – I saw her lying on the ground, wrapped in rags. A small child, a little younger than I was, perhaps eight years old, sat by her side. My first thought was: "For heaven's sake, *I* could be that child!" There was no difference whatever between the two of us but I lived in a house with a clean floor, my mother had a job, and this child looking at me lived in the dirt. Seeing that filled me with enormous gratitude for my undeserved luck and my privileged way of life. Nevertheless, we were not really well off at all. We lived exclusively off the income of my mother who clandestinely earned a little money on the side as a cabaret dancer. In the winter, I usually helped her to line her jacket with several layers of newspapers to keep her warm. This may show you our situation.

POERKSEN: Did your family always exist on he brink of poverty like that?

MATURANA: No. My mother's father came from a fairly well-to-do family in Bolivia. Having studied medicine in Chile, he went back there and was murdered. My mother was still very young when that happened – and because of that family tragedy she was taken to an Indian community in the Andes where she spent two years before she returned home again, penniless, to join her own relatives. Those two years were of tremendous influence both on her and on me because the Indian communities living in the Andes were not organised in a patriarchal-authoritarian way. Men and women live a balanced life in reciprocal supportive harmony borne by mutual respect. My mother told me that she, still a very young girl, learned there to appreciate a different culture of sharing and cooperation, in which all members of the community were involved according to their particular potential. She told me about that experience, and naturally it shaped my education. In retrospect, it seems to me that I actually grew up in a matristic family where self-confidence and self-respect were allowed to develop. My parents had separated soon after I had been born. My brother and I first lived with my grandmother who brought us up in the Catholic faith; after her death we lived exclusively with my mother. I would say today that my mother taught me what it means to accept responsibility and to act in a way that is both autonomous and respectful.

POERKSEN: How did your interest in the world of living beings develop? Did you too – as we can read in the biographies of other famous biologists – already run around with frogs in your pockets as a child?

MATURANA: More or less. To be precise, everything alive interested me for various reasons. One thing was that I was quite often ill, as I have already told you. Therefore, I already wanted to understand the meaning of death in early childhood. And so the thought naturally occurred that I had to try to understand the meaning of life as well because life and death were deeply interwoven and enmeshed. Another reason was that it gave me great joy to make things myself, to create something. At the age of eleven, I fell ill with tuberculosis and had to spend much time at home alone. I had some paper, scissors, and glue – and in many hours of work I created animals, cars, houses, a whole world. In this way, I built up a profound understanding of how the form of an entity – much later I would speak of the *structure of a system* – specifies and determines what operations may take place in it. What consequences, I asked myself, does the form have? After finishing school, I decided to study medicine because it was not possible, at the time, to take up biology. If you were interested in living systems, you had to choose either human or veterinary medicine. So in 1948 I matriculated at the medical faculty of the university but also prepared to study anthropology, ethnology and numerous other fields. Unfortunately, my university studies were interrupted for two whole years shortly afterwards because I had again become ill with tuberculosis. In 1950, after longer spells in hospitals and sanatoriums, I was finally dismissed as cured.

THE WARM-BLOODED DINOSAUR

POERKSEN: At some stage, I gather from a biographical summary, you left Chile to continue with your studies in England. There you encountered one of your important teachers, the neuroanatomist J. Z. Young.

MATURANA: In the year 1954 I was awarded a grant by the Rockefeller Foundation and took up work with Professor Young. He told me that I had to submit an essay on an agreed-upon topic of common interest every fortnight. One of the central rules he wanted to be kept unconditionally was that I provided an independent rationale for my arguments. Young taught me – like my mother earlier – to trust my own thinking. One day I offered him an essay in which I claimed that the dinosaurs had been warm-blooded animals. Some of my fellow students poked fun at my theory and called me the *warm-blooded dinosaur*. They thought that my view was an absurd heresy because the generally established opinion was then that only birds and mammals could have been warm-blooded but not the dinosaurs, that were classed as reptiles. Dinosaurs were, according to received opinion, reptiles and, therefore, cold-blooded. We now know that this is not the case at all. When I presented my arguments to Professor Young he was most interested – and sent me to a famous palaeontologist to discuss the theory of the warm-blooded dinosaur. In other words, he opened up for me a space of uninhibited thinking, of autonomous reflection. He always expected a serious and responsible debate, not the blind and thoughtless acceptance of some widespread opinion or mere academic doctrine.

POERKSEN: A few years later, you attained your doctorate in biology at *Harvard University* and then spent some time at one of the undisputed centres of the scientific world, the *Massachusetts Institute of Technology* (MIT). How did that come about?

MATURANA: There is a nice story about that. One day the distinguished neurophysiologist Jerry Lettvin was invited to speak at the customary midday meeting in the biological laboratory at *Harvard University*. He presented a theory of the process of vision. I spoke up, criticised him and invited him to my laboratory to demonstrate my own work. I was in the process of concluding my thesis on the anatomy of the optic nerve and the centre of vision in the brain of frogs. Lettvin was quite impressed and invited me to do post-doctoral work with him at MIT.

POERKSEN: Your divergent views did not cause the disruption of your contact but formed the basis for active cooperation.

MATURANA: Precisely. However, before we finally started working together and became friends, I asked him to give me some time to think about his offer, and tried to collect information about him from other people. What I heard about him at Harvard was rarely positive. Jerry Lettvin was erratic, people said, he would never finish his work and was a bit crazy. But I liked this tall man with his free spirit who was, at the same time, so warm-hearted and imaginative, and so I went to MIT in 1958. He, on the other hand, was enthusiastic about my thesis, showed it to everyone, and helped me to set up my own neuroanatomical laboratory, a small place just for me, for my own experiments. I usually worked there every morning until about one o'clock until Jerry came by and asked: "Humberto, who among our colleagues would be most irritated by the observations we made yesterday? Whom should we visit and provoke a little?" I quoted a few names and then accompanied this magnificent polemicist who never lost an intellectual dispute to the colleague we had chosen for the day. While Lettvin told him about our work, I was listening with gleeful pleasure. – It was a glorious time.

POERKSEN: I suppose that you must have visited the artificial intelligence star Marvin Minsky in his laboratory quite often. Seriously, though: As far as I know, Marvin Minsky was already at MIT, and his theories of human beings as "information processing systems" and the activity of thinking as some sort of "data processing" were probably not particularly attractive to you. They are in direct opposition to your view of communication, your description of structural determinism, and your characterisation of living systems. Did Minsky's work influence you in any way at all – perhaps as a kind of negative deterrent?

MATURANA: One might say so. When I went home in the evenings, I had to pass the entrance to the laboratory where the protagonists of artificial intelligence worked. I slowed down my steps and just listened to the conversations and discussions going on there. What I picked up in this way did not seem plausible to me at all. Marvin Minsky and his collaborators kept asserting that they were trying to create models of biological phenomena in their laboratory. That appeared completely absurd to me. What those people were doing was something entirely different, I thought. They were trying to create models of the surface appearance of a biological phenomenon without understanding what was going on inside the system that is responsible for the generation and production of a particular behavioural surface. I also did not like their extremely formalistic and mathematical approach. Whenever I appeared in one of those laboratories, I was showered with mathematical theories, arguments and formulae.

POERKSEN: What is it that sparks off this criticism of yours? Can mathematical reflections render the variety of biological phenomena invisible? Is there a kind of reductionism that you reject for aesthetic reasons?

MATURANA: No. I think that a formalism should be implemented only when complete understanding has been achieved of what the problem is and what is actually happening. If you use a formalism, you express your momentary understanding – and abstract from it. Some coherences have been grasped and understood, and, as a consequence, a formalism is constructed together with a network of relations that appears appropriate to the consequences of the coherences that have been understood. I would say, therefore, that my argument is not an aesthetic but rather an epistemological one. A formalism can interfere with the proper understanding of phenomenon and lead one astray. When a student in Chile asked me in 1960 what actually began four billion years ago that allows us to state today that life originated then, I did not want to make the same mistake and construct a model of the phenomenal image of a living system. What processes must take place to bring about the formation of something that we call a living system – that is the proper question to be answered.

WHAT THE FROG'S EYE TELLS THE FROG'S BRAIN

POERKSEN: What sort of research work did you do at MIT? What were the topics?

MATURANA: You ought to know that I like it very much when I can have my own room to do things that not everybody need be aware of. In October 1958, in my little laboratory at MIT, without telling anyone, I was studying the retinal cells of a frog – and I made a significant discovery. I could reveal under the microscope that there were evidently two fundamentally different types of cell. Some of them had fibres that radiated in a starlike pattern from the cell body and should therefore, I thought, be able to react to visual stimuli from all directions; the fibres of others extended long straight branches in one direction only so that a stimulus, I assumed, would correspondingly effect a unidirectional reaction. When Jerry Lettvin did not appear in his laboratory for five days I said to myself: This is my opportunity! Now I can test my hypothesis that the shape of the cells corresponds with its reaction. This was a completely new idea then, because in those years the process of vision was usually studied by projecting spots of light at the eye. The retina would, according to established opinion, receive the information coming in from the external world in the form of flashes of light and compute the corresponding reaction; that was the research dogma.

POERKSEN: Your observation of those special cells and their shape were a first step, I suppose, towards the epistemology you developed later: The structure of the organ of vision, and not the influence of an external world, is the cause of a particular perception.

MATURANA: Exactly. In the laboratory, I could not work with spots of light because I did not know how to use the apparatuses and was afraid I might damage something. I had to content myself with moving my hand in front of the frog's eye and simultaneously recording the impulses of an isolated cell in the optic nerve with an electrode. And I actually managed to discover a cell that reacted independent of the direction in which I moved my hand. Then I changed the position of the electrode slightly – and hit upon a cell that reacted only when I moved my hand in a particular direction. The cell showed the expected reaction in one direction only. I thought this discovery was fantastic; and I completed my experiments. When Jerry Lettvin returned two days later I told him about my discoveries. This wonderfully flexible man was fired with enthusiasm at once and said: "Now we shall do everything differently!" And he began immediately to re-arrange the whole laboratory so that our research and our discussions could be conducted in a completely new way. The experiments carried out there finally led to the publication of the two articles *What the frog's eye tells the frog's brain* and *Anatomy and physiology of vision in the frog*.

POERKSEN: The titles of these articles alone seem to indicate an epistemological tendency, which was to become even stronger in your later work: The exterior gradually loses in importance. The focus is no longer on the world telling the frog's eye about its properties, but centrally on the eye itself.

MATURANA: In these studies you can certainly see a step in this direction but not yet a fully thought-through re-orientation. It was not until 1965 when I – back in Chile – performed my experiments on the colour perception of pigeons that the critical transformation of my whole epistemology actually took place.

POERKSEN: At MIT you also met Warren McCulloch and Walter Pitts. Both were among the first cyberneticians in America, both were regular participants in the *Macy Conferences* that enabled cybernetic thinking to gain clear contours for the first time. At the centre of this kind of thinking are the figure of circular causality and the key example of the steering of a boat: The helmsman who wants to steer his boat safely into a harbour does not run off a program fixed once and for all but constantly varies his program. When the boat veers off course he assesses and corrects the deviation so as to continue moving towards the harbour. Correcting his mistake may produce oversteering and, consequently, a new deviation that, in turn, requires a re-adjustment of his course. Steering generates an effect that becomes the new cause for a new effect. And so on. What emerges is the figure of a causal circle, a circle similar to the format and the design of your own epistemology. Therefore the question: Did the encounters with cyberneticians like Warren McCulloch and Walter Pitts influence you in any way?

MATURANA: Not really. Of course, I met McCulloch from time to time but we did not do very much together. The relationship with mathematician Walter Pitts was more of a personal kind; he came to see me in my laboratory off and on, and I valued his sensitivity and tenderness and was touched by the fact that he went to Warren McCulloch's house every day to feed and assist Warren's mother, who was very old and fragile. That was marvellous. 99 percent of the time, however, I worked with Jerry Lettvin, who suggested one day that we consider both his mentor McCulloch and Walter Pitts as co-authors for the publication of our articles *"What the Frog's Eye Tells the Frog's Brain"* and *"Anatomy and Physiology of Vision in the Frog"*. Pitts needed publications and Warren McCulloch had taken on the role of his intellectual father. I accepted that. I was not, however, formed or influenced intellectually in any way by McCulloch or Pitts.

POERKSEN: Did you not find the confrontation with cybernetic thinking inspiring? While I was preparing myself for our conversations here, I noted the idea that you have given the cybernetic idea of circularity an epistemological turn and a philosophical foundation, that you now actually represent a *cybernetic epistemology.*

MATURANA: I did not encounter cybernetics proper until I met and became friends with Heinz von Foerster. At MIT, in those years, the concept of information was central, not the idea of circularity. – When Warren McCulloch states that the organism receives some *feedback* from its medium, I cannot see this as a perfect manifestation of circularity: If you describe organism and medium in this way, you have separated them from each other. A conception of this kind according to which the organism causes something to happen and then receives some feedback from its medium, is similar to moving to and fro between the two end poles of a linear relation. Strictly speaking, this is pseudo-circularity. Moreover, there is the additional assumption that the feedback contains some sort of message about the properties of the medium, which are, in this way, considered significant in themselves. Such a view is, as you know, completely alien to me.

POERKSEN: How would you describe the circular processes of knowing and living?

MATURANA: When I speak of circularity I refer to a circular dynamics of the organism (and that means: a circularity within the nervous system as well as a circularity in the realisation of autopoiesis) that makes the organism interact with the medium as a complete circular entity. Interacting with the medium does not disrupt the organism's circularity but leads to structural changes that in turn change the flow of circularity. This has nothing to do with a feedback from the medium or an input-output relation; it involves the reciprocal structural change of both organism and medium. This is a completely different situation. And when this circularity is destroyed by the encounter with the medium then the organism perishes.

Fig. 12: The circular worldview finds a symbolic expression in the figure of the Ouroboros, the snake eating its own tail.

2. Return to Chile

COMPETITION MEANS DEPENDENCY

POERKSEN: 1960 marks a break in your professional biography. You return to Chile, leaving the centre of Western science, although there were enough opportunities available to you to embark on a career in the arena of established American research institutions. The decision to leave the United States seems rather strange at first glance. Why did you leave MIT at all? The computer scientist Joseph Weizenbaum, who worked at MIT for nearly the whole of his professional life, once told me that he knew of people who would gladly give their right arm to be accepted by that institution. That is certainly a bloodspattered image to express the enormous magnetism of MIT. But you left and turned your back on North America. How did that come about?

MATURANA: There were several reasons for this decision. One was that in Chile I was protected against the tough competition ruling the science establishment. I am not at all a believer in competition who enjoys developing his ideas in opposition to others or presents them as a criticism of existing theories and concepts; I prefer an independent form of existence that does not restrict the freedom of reflection. If you do not have to compete you can fall back on your own special qualities and act according to your own standards and on the basis of your own responsibility. You are no longer dependent on whether you have published more articles than somebody else, whether you have advanced, have achieved a higher standing, have performed more experiments, but you are autonomous in your thinking and do not have to orientate yourself according to anyone's expectations. If you participate in competition, however, you subject yourself to the standards of somebody else's work and accept them as the relevant standards of quality for yourself.

POERKSEN: Competition is, if I follow you, actually dependency.

MATURANA: It is. You make yourself dependent; you lose your autonomy. For me, in those years, Chile was a competition-free zone. Another reason which contributed to my decision to return was that I felt responsible for my country that had given me so infinitely much ever since my childhood. When I was ill, people helped and cured me; when I went to school, I was allowed to learn although I had no money; when I studied at university, I did not have to pay anything.

POERKSEN: How did you experience this huge leap into a very different world? Did you never feel the desire to return to the North American world of science?

MATURANA: I was quite aware of the fact that I would not be able to work at the forefront of scientific research in Chile – and I asked myself what I should do. Was this the perfect occasion for falling into a depression? Should I change my profession in order to earn enough money outside the university to be able to live up to my expectations? Should I return to the US after all where I had been offered a professorship at Washington University in St. Louis? Or should I simply continue what I had started? I decided for the last option; I did not fall into a depression, I did not complain, and I did not return to the US. I stayed in Chile and at the university in order to carry on working in my own particular way.

POERKSEN: In your articles and books there is the odd remark to the effect that the research topics you chose often enough met with a hostile reaction from within the science establishment. When you were studying the colour perception of pigeons and began talking about the closure of the nervous system you probably did not endear yourself to your colleagues. At that early stage, you were yourself at heart still a realistically inclined scientist, I suppose.

MATURANA: That is right. As late as 1965, I wrote a short essay for a journal of the medical faculty where I was working as an assistant after my return, in which I declared that scientific activity rested on two fundamental assumptions. We must believe, first, that there is an observer-independent reality, and secondly, that our own statements refer to that reality in a recognisable way, even though we may never be able to grasp this reality completely. A few months after the publication of this little article, however, my views had changed entirely. I discovered that it was impossible to find unambiguous correlations between a physically specified colour and the activities of retinal ganglion cells of pigeons. And when I made these findings known and discussed them with my colleagues many members of the university said I had gone crazy.

POERKSEN: I am told that you were called to the director of the medical faculty one day to be told that your research had nothing to do with reality. Obviously there were rumours that a highly gifted young scientist had regrettably strayed from the path of acceptable science. Did the episode actually occur like this?

MATURANA: On the whole, yes. People said I was talented but sterile, just not creative. They said I should have avoided the topic of cognition and should simply have continued with my experiments – and I would have already been awarded a Nobel Prize. I asked: "Are you implying that I ought to leave the medical faculty?" The answer was affirmative. Naturally, I was hurt by the thought that my work was obviously not appreciated, but one day a friend of mine asked me whether I really had to be understood, whether it was at all necessary for me to be understood by other people. He is actually quite right, I thought. Why does one have to be understood at all? What was most important to me was to carry on with my work in a serious way. In the intervening period I did not avoid any debate and defended my views unswervingly. Perhaps people did think me crazy but that did not really affect me particularly or put me under special pressure. My claims had not yet been refuted.

POERKSEN: Studying the history of science brings to light diverse cases of truth terrorism that have often ruined people completely. It is possible, of course, that you too might have ended your career unknown somewhere in the periphery of research, and that nobody would ever have heard anything about your work on the colour perception of pigeons or the biology of cognition.

MATURANA: It is conceivable but it did not happen this way. In the early sixties, great efforts were undertaken in Santiago to establish a centre for the training of young scientists at the University of Chile. I took part in the setting up of a faculty of sciences, became one of the teachers, and was finally appointed professor.

INSIGHTS OF AN OUTSIDER

POERKSEN: Going by the chronology of things for a moment, we are approaching the end of the sixties: the period of unrest from Berkeley to Paris. The Vietnam War broke out, the student protests started. How did you experience this phase of history in Chile?

MATURANA: I joined the protests in my country – despite massive reproaches from my colleagues, and despite my status as a university assistant. The demonstrations had begun at the Catholic University and spread continually. One day the students had occupied the medical faculty. Finding myself confronted by them, I asked them to feed the animals in my laboratory and then to allow me to join their meeting. It soon became clear to me during those meetings, which were supposed to deal with the future of the university, that nobody had a precise conception of what should actually be done. So I finally stood up and suggested a three-stage debate about university education: A first day was to be devoted exclusively to criticism and to finish with a plenary meeting assessing the results; a second day should deal with the students' wishes and goals; and a third day should discuss their possible implementation. The professors denounced me as a political agitator; the students were jubilant and thought I was one of them. For three days, everyone was listening to everyone, common plans were developed in a spirit of both seriousness and enjoyment, and cooperation developed that was to last a whole month. That was a fantastic experience because the political clichés – *He's a communist! He's a liberal!* – gradually dissolved. That period taught me how to act through listening, how listening changes in the course of different meetings, and when the right moment has arrived in a discussion to speak up oneself.

POERKSEN: Every now and then, you met people who were prominent during the rebellion of the sixties and seventies, or who belonged to the avant-garde of a new kind of thinking. The culture-critic Ivan Illich invited you to Cuernavaca in Mexico, the modern Zen mystic and psychotechnologist Werner Erhard asked you to California. And you taught at the Naropa Institute of the Tibetan teacher Chögyam Trungpa in Boulder, Colorado. Would you say that the intellectual climate of the sixties and seventies – that vehement search for autonomy in the political and the private sphere – made an impression on you? Or is this nothing but an accidental collection of occasions for public lecturing?

MATURANA: I would say that we are looking at a number of accidental opportunities and invitations offered to me, sometimes through friends of mine. The experiences were definitely not particularly penetrating. The Naropa Institute did ask me to give a seminar but kept a safe distance from my ideas: the focus there was, naturally, on Buddhist and Tibetan psychology. When Werner Erhard invited me, I had to familiarise a relatively small circle of his collaborators with the biology of cognition and then attend one of his coaching sessions and write a report about it – which I did. The time spent with Ivan Illich in Cuernavaca was not formative for me, either.

POERKSEN: Why not? That seems rather unusual to me.

MATURANA: You must realise that never in my whole life have I ever been a member of a group or a political party. At the age of eleven, I left the Catholic Church because I had – in view of all the pain and suffering – begun to consider God unjust. How could an almighty, omniscient and benevolent God allow all the innumerable injustices that I witnessed? His benevolence, so I found, contradicted his omnipotence and omniscience. Since I left the Church as a little boy, I have never again associated myself with a particular religion. I never belonged to Werner Erhard's organisation, or the Tibetan group at Boulder. In addition, I do not consider myself a Buddhist or a believer in the ideas of Ivan Illich. I do not intend this as a criticism or as some kind of depreciation; in a certain way, I have always remained an outsider.

POERKSEN: You were there but merely as an observer.

MATURANA: I would rather describe myself as a sort of parasite. I was in place, I listened, I did my thing, but I was not part of the organisation or religion. Insiders, however, get friendly with all the important people, adopt their view of the world, join the relevant group or party whose goals they will then promote.

POERKSEN: But are not insiders the happier ones? The outsiders' life is necessarily a lonely one. They are homeless.

MATURANA: Not necessarily, because they may find their home in themselves.

POERKSEN: What would you call this home?

MATURANA: Autonomy, self-respect.

POERKSEN: What are the advantages that outsiders enjoy? That you cannot hurt them?

MATURANA: I would say so. They can lead their lives the way they want to, free of any pressure to defend particular principles. They do not feel committed to any ideology and are free to enjoy all the opportunities for reflection. Outsiders participate without prejudice and can, therefore, perceive what appears before them. All this gives them an advantage over insiders.

POERKSEN: Is the position you describe only an accidental predilection or indeed more? Is it not the expression of the living incorporation of a theory? A submerged current of your thinking seems to be keeping actuality at a distance. You describe – without direct involvement, without concrete entanglement – the conditions of the possibility underlying all knowledge.

MATURANA: That is correct. Those who act in this role of the observing outsider should be capable of a triple look of the most unprejudiced kind. They must be able to look inside the system and identify its components and their reciprocal relations – and still remain aware

Fig. 13: "I am thanking the pigeons that I experimented with in my laboratory. It was a sort of ceremony; it helped me to preserve awareness for what I actually did. There was no transcendental justification for destroying these animals, – truth, scientific progress, the prosperity of humankind, or anything similar. What I did to the pigeons – in order to understand the nervous system – is entirely my own responsibility."

of how the system appears as a whole in the domain of interactions, and how this domain is in turn related to the domain of internal relations in a meta-domain. What does one see if one observes in this way? Naturally, one does not recognise some objectively given reality, that is obvious, but one does indeed achieve an adequate understanding.

POERKSEN: Some people might see this distanced view of the observer as a form of indifference.

MATURANA: These people stick an emotionally coloured label on such an attitude. They attack indifference and simultaneously demand involvement. In my view, observers practise a form of participation that can neither be classified as indifferent nor as involved. The decisive thing is that observers must not be influenced by their own ambitions and the wish to achieve a specific result. This is precisely the reason why an observer can perceive anything at all; for whoever wants to see and understand something, must let it happen and let it show itself. The motto for the kind of perception that makes such understanding possible, and is founded on love, is: *Let it be!*

POERKSEN: Could you present an example from your everyday research work to clarify this attitude of the outsider?

MATURANA: I shall tell you a little story. One day I decided to learn how to fly because I was investigating the processes of vision in pigeons in my laboratory and wanted to understand how these birds experienced the world in their very own element. When I appeared at the school for glider pilots and started my training, I was in the role of the outsider again who was unwilling to take part in the usual exchanges on the airfield. My goal appeared extraordinary, too, and strangely bizarre. – Who would want to understand a bird?

THE *TRACTATUS BIOLOGICO-PHILOSOPHICUS*

POERKSEN: In 1968 you left Chile once more to work with biophysicist and cybernetician Heinz von Foerster at his *Biological Computer Laboratory* (BCL) for ten months. The BCL at the University of Illinois was a small interdisciplinary republic of scholars, at the time: neurobiologists, electrical engineers and dolphin specialists worked together with philosophers, physicists and logicians, and many of the research projects carried out there have proved to be groundbreaking and formative with regard to the style of epistemological discussion ever since. The paper you are most famous for, entitled

Biology of Cognition, first appeared as a research report of the BCL. How did that text come to be written?

MATURANA: A few weeks after my arrival in November 1968 Heinz von Foerster asked me to prepare a paper for a conference under the heading *Cognition: a Multiple View* that was to take place in Chicago. Anthropologists were also to attend that conference which was organised by the *Wenner Gren Foundation*. My task was to present the neurophysiology of cognition. My first thought was that all these people would politely listen to me speaking about nerve impulses, synapses, etc. but then pass on to other topics and promptly forget what I had been saying. However, I did not want to be forgotten. I worked out a more generally accessible synthesis of my view of the nervous system and cognition and spoke about the observer.

POERKSEN: "Anything said," we read in the paper published later, "is said by an observer."

MATURANA: That very sentence I wrote on the blackboard during my lecture – and from that moment, the observer was present in all the talks that took place. As I had decided to speak about the process of knowing the knower as the fundamental condition of all the processes involved inevitably moved to the forefront. Whatever is said, I wanted to emphasise, can under no condition be separated from the human being saying it. There is no division between the speaker and what is spoken. The observer is necessarily the origin and source of everything. For the anthropologists present at the conference this was a fundamental insight.

POERKSEN: How did *Biology of Cognition* come to be published?

MATURANA: When I had returned to the BCL I reworked my conference text and gave the new version to Heinz von Foerster with the request to correct my English – he called it *Spanglish* – together with a student and to delete all the repetitions. When I got the paper back, I was furious. I thought that my text had been destroyed. Heinz von Foerster said he had only cut all the repetitions; in my view, however, he had linearised my method of circular discourse.

POERKSEN: I assume that it is generally difficult for you to write short articles because the brevity of a text does not permit a comprehensive presentation. The circular process of knowledge creation is, therefore, unavoidably cut short at a certain point.

MATURANA: I am aware of this problem, too. Usually we speak and write about things as if they had an observer-independent existence, but that is precisely what I do not want to do – and so I try to speak and write in a manner that shows that nothing exists independently or could be separated from an observer. It is very difficult to generate awareness for the processes of the production of what we commonly consider given in the process of writing itself.

POERKSEN: This would imply that a new kind of thinking might possibly require a new kind of speaking and writing. However, there is another problem. If you want to arouse and heighten the sensitivity for the circularity of all thinking you simply need time. The firmly anchored realism of everyday life must be transformed gradually into a different kind of worldview that may in turn lead to new modes of experience. Undoubtedly this will consume time and energy. Is such persuasion work not fatiguing in a world bent on quick understanding?

MATURANA: That is not my problem. I do not want to convince anybody, or convert people to a circular worldview. I am not a revolutionary and I do not see myself as a man with a mission to change the world; I simply want to demonstrate how certain processes produce certain entities, that is all. I am living today as if I had infinite amounts of time, without hurry or haste, moving at my own speed. In earlier years, things were possibly different – even in the early sixties. I wanted to convince people of my views. I am completely cured of such intentions now, because a friend had said to me one day: *The harder you try to convince people the less trustworthy you become.* I think he is right.

POERKSEN: Looking back, you wrote about the time you spent with Heinz von Foerster at the BCL: "Maybe we did not work in the usual sense, but we talked a lot and embraced each other frequently, spending many full hours conceiving a Tractatus biologico-philo-

Fig. 14: Heinz von Foerster and Humberto R. Maturana on the fringe of a conference.

sophicus that we never had the time to write." How did you meet Heinz von Foerster, in the first place? How did the contact come about?

MATURANA: I did not meet him through some complex intellectual debate but by way of a playful and exhilarating encounter on the fringe of a conference of physiologists at Leiden in Holland. When the Dutch queen began to thank the organiser of the conference during the opening ceremony, we both fled, more or less at the same time, and we ran into each other on our way out, confessed to each other that we did not like ceremonies, and decided to go to Amsterdam to stroll through the museums. It was a wonderful excursion and we laughed a lot together and enjoyed ourselves like two old playmates.

POERKSEN: What form did your collaboration take at the BCL?

MATURANA: Heinz von Foerster used to work through the night into the small hours of the morning and therefore rarely appeared in the laboratory before midday. Often he came right to my room and we talked for a while. I took part in the heuristics seminar that he conducted together with Herbert Brün where I had, it seems to me, the function of a strange oracle that rarely spoke. From time to time I said something about the observer or about the double look with which one may approach a system, and then everybody was silent until the discussion got going again. During my time at the BCL, I worked with several students, occasionally talked to the cybernetician Ross Ashby or the philosopher Gotthard Günther, who taught there in those months, worked on the text of my *Biology of Cognition,* and regularly visited Heinz von Foerster in his laboratory or in his house in Illinois.

SYSTEMIC WISDOM

POERKSEN: In a festschrift you once described him as a Zen master in the art of handling systems. How are we to interpret this?

MATURANA: Heinz von Foerster has a most profound understanding of systems. He recognises their matrix and spots the gaps and empty spaces that are not covered by this matrix. In these gaps he moves with complete freedom and perfect self-confidence and is, should the need arise, able to make himself invisible. I remember travelling to town with him one day to get a couple of things done and looking for a parking spot. Heinz von Foerster parked the car in front of the police station right under the notice: "Parking only with special permission." He got out of the car quite confidently, and I asked him nervously why he had chosen precisely this place to park and whether he really had special permission. "No," he said, "but as everybody is aware that you can park here only with special permission even the police will believe that I obviously have it. Otherwise I would never dare to park my car in this place!" – "My goodness," my reaction was, "I would give myself away at once!" – "I know", he said, "because you actually believe that you have no right to park your car right here." This dialogue was most illuminating for me because it revealed the systemic understanding Heinz von Foerster commanded, – and, at the same time, my lack of self-confidence. Whoever wants to act in a system, I realised, must not only understand it but also fully trust this understanding and act accordingly.

POERKSEN: After the months spent at the BCL you returned to Chile, with the first written synthesis of your theory of cognition in your bags. Francisco Varela worked with you there, with whom you published a number of books, amongst others the bestseller *The Tree of Knowledge*.

MATURANA: When I was back in Santiago I helped Francisco Varela, who had earned his doctorate in Harvard, when he came back to Chile, by making room for him in my laboratory. If my ideas about the circular organisation of living systems were adequate, he said to me one day, then it should be possible to formalise them. I said that a full linguistic description was needed before any formalisation could be sensibly attempted, because only what is completely comprehended should be expressed by an adequate formalism.

POERKSEN: This means that the criterion for the introduction of a formalism is the point in time at which one begins developing and applying it. Premature formalisation may deprive one of comprehensive understanding and block one's thinking.

MATURANA: Exactly. You do not translate the actual phenomenon into a formalism but the momentary understanding of the phenomenon as far as you have been able to develop it. It has therefore always been most important for me to begin with a linguistic description, and so we wrote and published *De Maquinas y Seres Vivos*, a little book about machines and living systems.

POERKSEN: Francisco Varela positions your cooperative theoretical reflections devoted to the organisation of the living, which finally resulted in the theory of autopoiesis, in the context of the political climate of Chile. The communist Salvador Allende was elected president, and those who wished to do so were inclined to see signs of the beginning of a new era: "It was clear to us", Varela writes in retrospect, "that we were about to embark on a journey that was decidedly revolutionary and unorthodox, and that the necessary courage to do so derived from the prevalent mood in Chile … The months leading up to the emergence of the concept of autopoiesis are inextricably linked to the Chile of those days."

MATURANA: I would quite definitely contradict that. I am not at all interested in occupying a revolutionary or unorthodox position and, consequently, in assessing my work by corresponding standards. Perhaps some of my ideas will seem revolutionary to some people but I myself was never a revolutionary. All I want is to do my job properly. What Francisco Varela writes about that time is his personal opinion. He was just beginning to make himself familiar with my deliberations concerning the organisation of the living; he was my student discovering and learning to understand something that had occupied me long before, in fact since the days of my childhood. This is not meant to sound aggressive but the fact is that I had already developed all the concepts when we started working together in our laboratory and writing our papers and books in 1970. Once again: My conceptions involving the autopoiesis of living systems had nothing whatever to do with what went on in Chile at the time. More exactly, it was the other way round: I was able to make good use of my theoretical ideas to understand what was going on in my country.

POERKSEN: Could you report an example?

MATURANA: A short while before Allende was elected, I attended, for curiosity's sake, a meeting of a political group calling itself *La O* (The Organisation), together with Francisco Varela and our mutual friend José Maria Bulnes. A communist founded it, and its goal was to enlighten the workers in the factories about the privileges and the salaries enjoyed by only a few people. For this purpose, a sort of mini-newspaper was created which we secretly distributed among the workers during the night in order to enable them to observe their own living circumstances. When Allende was finally elected, it was generally said that the Left had now gained democratic access to power. The members of this group now met for consultation. Should they, the question was, disband? Should they continue working in the underground? Would it not be sensible to integrate the group in one of the established parties? Although I did not belong to the circle of those who made the decisions, I managed to take part in that meeting. And at some stage of the proceedings I intervened and said: "You are making a mistake. You are talking as if Allende were an elected president but that is false. The fact is that Allende was appointed president – but that is another matter. Of the three candidates he had only a tiny minority of votes."

POERKSEN: Allende won a good third of the votes.

MATURANA: That is correct. And two thirds of the population had not voted for him. His relative numerical majority did not mean, I insisted, that the majority of the Chileans had voted for him and would now support him. I therefore demanded: "In such a situation, your organisation, the disbanding of which you are debating here, should try to gain more power and, in any case, continue working in the underground. The real challenge is still to come." The group, of course, disbanded – and one day the opposition had attained such strength in the country that the putsch could be ventured and everything was over. Even today, this discussion appears to me to be a kind of exemplary lesson: These people were blind to the dynamics that had swept them into their present situation. They lacked the ability of observing. That was a very important experience, because I had found myself confronted with my own *theory in action*. But the fundamental ideas that finally led to the concept of autopoiesis had been developed much earlier.

POERKSEN: Could the rift that I can sense between you and Francisco Varela have to do with different styles of thinking? Varela is very keen to translate ideas into a mathematical language, to formalise them, whereas you have always been very critical of such an interest in formalisation, – also in our conversation here.

MATURANA: This is definitely a crucial point. I have always been a biologist but he has been, I would say, more of a mathematician.

THE BRAIN OF A COUNTRY

POERKSEN: You have expressed very clearly that you actually resent the daily business of politics – the rhetoric of breaking new ground, the idea of changing the world, the fundamental element of a mission, etc. In spite of that, your ideas have undoubtedly had some political influence, if I am correctly informed. Under Allende, the Chilean Fernando Flores, 26 years young, was made minister of both economy and finance, and finally spokesman of the government. He invited the cybernetician and management consultant Stafford Beer to Santiago and together with him designed Project *Cybersyn*. It was intended to create a centralised model to plan and control industrial production. It was to operate as an early warning system so that changes in production could be recognised in good time and then met by appropriate measures. Stafford Beer – who wrote an introduction to one of your papers – wanted to understand the whole economic system as a sort of nervous system, and to create a central *observation room* where all the economically relevant changes were to be registered. Would you say that Stafford Beer and Fernando Flores were influenced by your ideas?

MATURANA: No, I do not think one can say it in this way. Fernando Flores was strongly affected by Stafford Beer's book *The Brain of the Firm*. When Beer came to Chile for the first time in 1972, he asked to see the Chilean cybernetician Humberto Maturana. Nobody had the slightest idea who that Maturana might be whom the great Stafford Beer wanted to meet. Finally, I was tracked down and invited to a meeting.

POERKSEN: The system of programs that was built up provoked violent criticism because it was considered an early dream of socialist planning and control. The conception of this cybernetically inspired information system was obviously rigidly centralist, and, in the end, it was used to neutralize a lorry drivers' strike: newly built lorries were delivered ahead of time and manned by students as strike breakers.

MATURANA: That was not the goal of this project, and certainly not Stafford Beer's intention. It was Fernando Flores who wanted to implement *The Brain of the Firm* at the level of an entire country. And he invited Stafford Beer to help the engineers with its realisation and to teach them the necessary cybernetics. It is true that through his active involvement, production was, in fact, monitored in many places of the country in real time and the resulting data were collected in a so-called control room. In this room, developments in their initial stages were to be extrapolated with the help of suitable models in order to be able to take appropriate decisions immediately and modify action programs efficiently at the very moment at which difficulties actually arose or certain changes occurred, and not only months later. That was Stafford Beer's idea: a centralist system of management, but not an instrument of domination. The idea of control was not of central importance to him, it was perhaps to Fernando Flores. However, this central control room was, as Heinz von Foerster said, who was visiting Chile at the same time, not a real control room at all because the required data processing capacities and the sufficiently complex simulation models for experimentally testing the relevant different variations of particular situations were not available.

3. Experience of a dictatorship

THE EMERGENCE OF BLIND SPOTS

POERKSEN: Project *Cybersyn* and the plans of the socialist Salvador Allende met with a violent end on 11 September 1973. At two o'clock in the afternoon, troops of the putsching general Pinochet stormed the presidential palace – and at the end of the day Salvador Allende was dead, Fernando Flores was taken prisoner near the island of Sierra del Fuego, and General Pinochet ruled the country as dictator for many years. Many members of the university fled to other countries and emigrated to the US or to Europe. What did you do?

MATURANA: On the day of the military coup I rang my friend Heinz von Foerster and asked him to help me and my family to leave the country. The situation was dangerous: Many people were suddenly among the persecuted, there were dead people lying in the streets, there was a curfew, and there were arrests. Soldiers appeared at the university. Heinz von Foerster tried to get me an invitation from an American university, which was not at all easy, of course. I was considered a dissident in science who spoke of the closure of the nervous system although everybody knew that it was demonstrably an open system. I was known, but I did not belong to mainstream science. It was therefore not surprising that nobody wanted to have me at first, despite the efforts of Heinz von Foerster. The University of Illinois was not interested either. Ten days later, a neurophysiologist in New York had been found who was interested in my work. But by that time I had already decided to stay in Chile.

POERKSEN: How did you reach that decision? There was an exodus of the intelligentsia in those years, an escape from repression and torture. Tens of thousands of Chileans emigrated, and the opposition was the object of unremitting persecution that some 3,000 people did not survive.

MATURANA: My motives to stay were of different kinds. My first thought was: If all democratically minded people left the country there would soon be no recollection of a democratic culture and of another, better time. In this perspective, every older person was a living treasure. Then I was concerned about the fate of all the many students who were dispirited and suddenly found themselves drifting through the university on their own. Many professors had fled or gone into hiding, or had already been arrested. I met with some of them at the university one day, and we formed a sort of pact and decided to stay in Chile. I kept that pact and continued to work as a democratically minded member of my university because I felt responsible for the students and my country.

POERKSEN: You wrote once that one of your motives was to comprehend the essence of dictatorship.

MATURANA: That is true although it may sound a bit crazy. But I really wanted to know what it means to live under a dictator. I wanted to understand the Germans and, in particular, the history of my friend Heinz von Foerster who had survived the Nazi terror due to his understanding of systems. He once said to me: *The more specified a system is, the easier it is to cheat it.* I also asked myself whether I might be able to observe in such a dictatorial system how people gradually go blind, and what the causes of such perceptual deprivation were. Can one, if one has been duly forewarned and is aware of the dangers of ideologically produced blindness, prevent it from developing and retain one's capabilities of vision and perception? One of the goals of a dictator is always to deprive people of any opportunity to remain or become observers of their own circumstances, and to deny them all chances of changing these circumstances and transforming them according to their own desires.

POERKSEN: You wanted to come to grips with the epistemology of ideologies.

MATURANA: You might put it that way, yes. – When innumerable Germans insisted after the War that they had known nothing about the horrors of the Nazi period, I was convinced that not all of them were liars. Perhaps some of them were simply unable to face up to the terrible truth. I wanted to know what had been going on inside them and in their psyches. How does one live under a dictatorial regime that makes it so very difficult to keep away from it? In what measure does one unevitably go blind even though one definitely does not want it to happen? Does one go blind because one knows that one could? How and in what ways is blindness produced at all?

POERKSEN: What did you observe?

MATURANA: Nobody is everywhere. If you decree curfews, you prevent people from seeing certain things. They will be unable to notice that people are murdered in their street during the night; they will not see the corpses. Everything happens behind a curtain. So people might not believe the rumours and tales they come across when they go out in the morning. There is nothing to be seen, not even a trace of blood, and what has happened is strictly denied and rejected by the authorities. Moreover, people will probably say to themselves that soldiers are human beings too, and that no human being can behave in such bestial ways. Such humanist presumptions may therefore very well make us blind: they protect us against the horror and they preserve our trust in other people. Of course, the new situation of a dictatorship creates new advantages for some people: Suddenly particular jobs are available because other people had to give them up and get away.

IDEOLOGY AND THE MILITARY

POERKSEN: Comparing the Chilean and the Nazi dictatorships, we discover an essential difference: Adolf Hitler created an ideological dictatorship. He tried to win elections, although by applying massive means of intimidation, on the one hand, and he wanted to convince and excite the masses about his mad anti-Semitic ideas and the religion of racism, on the other. The military dictatorship in Chile was primarily based on the force of arms and the power of the army; its ideological underpinnings were rather weak.

MATURANA: That is a central point. The mental freedom of movement of people living under an ideological dictatorship is doubly limited: it is decreed, on the one hand, what has to be believed, and it is specified, on the other, what must never be said or thought if any risk of endangering life or status is to be avoided. A military dictatorship primarily lays down what must not be done. In the Chile of those years, any kind of criticism of the government and any sort of support for the ideals of socialism was prohibited. Apart from that, you could think and teach whatever you liked.

POERKSEN: Pinochet kept reiterating that the Left was against the family, private property, freedom and the fatherland. He used just a few meagre ideological phrases, nothing more.

MATURANA: It was an *anti-ideology*, directed against communism. We were, after all, Pinochet kept pointing out all the time, in a state of war, and in a state of war you have to kill your enemies, – that was his argument. He used this state of war declared by himself to justify the violations of human rights that were committed.

POERKSEN: A central element of the Chilean dictatorial regime was the *miedo,* the terror, the spreading of fear. The singer and guitarist Victor Jarra was arrested, had his hands smashed, and was finally murdered. The poet Pablo Neruda was isolated; his houses were searched. People were tortured. Did people know about all this?

MATURANA: Yes. For over a year every television newscast had to start with the bombardment of the government palace, then came reports about the arrest of *revolutionaries* and the discovery of secret arms caches. And so on. We should, however, not forget that Pinochet had the support of a significant majority of the population. Many people acquired enormous wealth under his regime through the privatisation of public property and, therefore, profited directly from the activities of his government.

POERKSEN: I find it striking that you and various other authors, who are counted among the founders of constructivism today, all had to suffer under dictatorial regimes and were confronted with dogmatic worldviews. Heinz von Foerster had to hide from the Nazi thugs;

Ernst von Glasersfeld left Vienna when the Nazis seized power; Paul Watzlawick has repeatedly suggested how deeply shocked he was by the Nazi regime; Francisco Varela escaped from Pinochet to Costa Rica, and you lived in Chile all those years. My question is now: Is there a connection between the theories of these authors and the experience of dictatorship? Alternatively, is this biographical correspondence purely accidental?

MATURANA: It is not accidental but the result of the period. Many people were confronted with authoritarian systems more or less directly during the past century –the century of the Russian Revolution, of Fascism and Nazism. I can, of course, only speak for myself, but my own understanding of power does not derive from the experiences I went through after the military coup in Chile. Rather the reverse. My life under the dictatorship was informed by my understanding of power, resulting from my permanent longing for democracy. Supporting democracy obviously entails the rejection of dictatorship that, therefore, becomes an enemy and a constant threat lurking in the background. All those actively engaged in the democratisation of a country quickly realise how difficult and laborious it is to keep a democratic culture alive. One has to come to terms with the ideal of perfection, which is widespread and deep-rooted in our culture, and with the attempt to generate seemingly perfect and allegedly democratic forms of living together even with the means of oppression. One is evidently opposed to dictatorship and, consequently, an active supporter of the individual, not of the goals of some collective. Still, one must not lose sight of the whole of society when working for the democratic participation of the individual. The people you mentioned have, I think, been well aware of these difficulties and understood that there is no antagonism between individual and society. This is what they all have in common.

THE POWERLESSNESS OF POWER

POERKSEN: Your contributions to systems theory and the biology of cognition always deal with the autonomy of individuals and their particular ways of looking at the world and moving around in it. You claim that all human beings follow their own laws in cognition and action, that they are structure-determined systems. This conception sets narrow limits for the concept of direct and linear control. However, is not the wielding of power and force by dictators a compelling example of how extensively people can be controlled and influenced by external forces, after all?

MATURANA: No, that is not the case. As I have lived under a dictatorial regime, I know what I am talking about. Strangely enough, power arises only when there is obedience. It is the consequence of an act of submission depending on the decisions and the structure of the individuals subjecting themselves. It is granted to dictators by doing what they want. You grant power to others in order to keep or save something – life, freedom, possessions, jobs, a relationship, etc. I claim: *Power arises through submission.* When dictators or other people point a gun at me and want to force me to do something, then I am the one who has to consider: Do I want to grant power to these people? – Perhaps it is sensible to meet their demands for some time in order to be able to get the better of them in favourable circumstances.

POERKSEN: Does what you are saying also apply to the dictatorship of the Nazis? Was it the terror of the Gestapo that made Adolf Hitler powerful? Or did the people actually decide to grant power to a third-class painter from Austria?

MATURANA: It was a conscious or a subconscious decision of the people which gave power to Adolf Hitler. All those who did not protest had decided not to protest. They had decided to subject themselves. Suppose a dictator comes along and kills every person refusing to obey him. Suppose the people of the country all refuse to obey him. The consequence: He kills and kills. But for how long? Well, in the extreme case he will go on killing until everybody is dead. Where is the dictator's power then? – It has vanished.

POERKSEN: How do you want us to interpret this re-formulation of the relationship between power and helplessness? Is this an idealistic call not to subject ourselves? Or do you really mean what you are saying?

MATURANA: I am totally serious when I say: We always do what we want to do, even though we may claim to be acting against our will or to have been compelled to do something. In such cases we desire the consequences of our actions although we may not like what we are doing at the moment.

POERKSEN: Could you illustrate these ideas with an example?

MATURANA: Nobody can force you to shoot at another person but you may, of course, decide to shoot in order to save your own life. Maintaining that you were forced to shoot is only an excuse that obscures the goal you were pursuing, namely, to save your life for the price of subjecting yourself. When you decide, in such a situation, not to shoot at another person, a shot may still be heard but it will be a shot fired at you – and you might die, preserving your dignity.

POERKSEN: Would you say, therefore, that there are no real victims?

MATURANA: Strictly speaking, yes. Victims despise themselves because they have granted power to others and denied themselves in their autonomy by an act of obedience. In the self-description as victims, the actual processes of the generation of power are made invisible.

POERKSEN: The Chilean dictator Pinochet ordered, as we all know, the abduction, torture and murder of many of his opponents. What did you do when Salvador Allende was dead and the socialist experiment had met with a bloody end?

MATURANA: I decided to pretend in order to stay alive and to protect my family and children. At the same time, I tried to move and behave in such a way as to avoid endangering my dignity and my self-respect. I kept away from certain situations, respected the curfew, did not discuss certain topics in the university. – When the soldiers came and ordered me to raise my hands and to move up to the wall, I raised my hands and moved up to the wall. However, it was quite clear to me in those moments that the time would come when I would no longer be prepared to grant power to the dictator's regime.

POERKSEN: Would you like to tell me about a particular situation?

MATURANA: One day in 1977 I was arrested and put into prison. The reason was that I had given three lectures. The first lecture dealt with Genesis and the Fall. I said that Eve, who had eaten from the apple and then given it to Adam, could serve as an example. She was disobedient, and her rebellion against the divine commandment laid the foundation for human self-knowledge and responsible action, for the expulsion from paradise, a world without self-knowledge. In the second lecture, I spoke about St. Francis of Assisi. His way of perceiving human beings generates such deep respect towards them that it becomes impossible to define them as enemies. And I added that every army must first transform other human beings into strangers and then into enemies in order to be able to maltreat and kill them. The third lecture was devoted to Jesus and the New Testament. How do we live together, I asked my audience, if we base everything on the emotion of love?

POERKSEN: What exactly happened after your last lecture?

MATURANA: A few days later, I was taken to prison and treated like a prisoner. I was to be interrogated, I heard. One day somebody arrived and called out my name and said: "Are you Professor Humberto Maturana?" When I heard that I thought that I would remain a professor forever even if these people killed me. The status of professor was the protective shield they had granted me. They took me to a room where three people were waiting. I sat down and asked the question: "In what way have I violated the statement of principles issued by the military government?" This means that it was me who began the interrogation and changed the rules of the game. I would not say that I manipulated those people but that the interrogation took place in a way that allowed me to keep my dignity and self-respect. I continued behaving like a professor and tried to counter the accusations they formulated. And I gave these people a lecture on evolutionary theory and explained to them why they would never be able to destroy communism by persecuting communists. It was necessary to change or eliminate the conditions that made communism possible, in the first place. The three men listened to me with growing astonishment. I told them they could invite me for a lecture any time. Then they took me back to the university.

The Maintenance of Self-Respect

POERKSEN: Your very own experiences during the years of the dictatorship are most important to me because they make me understand you better, I believe. You do not plead for some fatal heroism, you do not condemn those who subject themselves, but you plead for a maximum of awareness in the handling of power.

MATURANA: Naturally, yes. It can be very stupid not to subject oneself for a time and to wait for a suitable opportunity to strike back. My fundamental point is to declare one's responsibility and to invite others to act in full awareness. Does one want the world that emerges when one grants power to others? Does one primarily want to survive? Does one reject the world emerging through the wielding of power in an unconditional and uncompromising way?

POERKSEN: Do you believe that that different state of awareness is decisive? It might be argued that conscious or subconscious subjection leads to the same consequences: the dictator stays in power.

MATURANA: This different state of awareness is decisive because it allows you to be hypocritical. Being hypocritical means simulating a non-existent emotion. You remain an observer, keeping an inner distance, and one day you may act in a different way again. This means that the perceptual abilities of the pretenders are not destroyed, and their self-respect and dignity are preserved. Due to these decisive and very significant experiences, they may be able to lead a different life. If one gives up this attitude of the conscious handling of power, one is lost because one has decided for blindness.

POERKSEN: How can we be sure that the belief that we are merely hypocritical and observing is not just a subtle and refined form of self-delusion?

MATURANA: Well, that is a difficult problem, indeed. The situation is particularly precarious when people are convinced that they are immune to the temptations of power. These people have become blind to their own temptability, to the delights of wielding power, the pleasures of the uncontrolled execution of control. My view is that we should never believe that we are in any way special as far as morality or anything else is concerned: we are then mentally unprepared for situations that may make torturers of us. Those who think they are immune will be the first, I believe, to become torturers in certain situations. They are not aware of their own seducibility. Whatever horrible or wonderful things one human being can do – there will always be another, and it could be you or me, who is capable of doing the same. Such an insight allows us to lead our lives in full awareness and to decide whether to support democracy or a dictatorship.

POERKSEN: Throughout the 17 years of the dictatorial regime in Chile, you worked as an academic teacher together with your students. How openly could you operate within the university? How did you conduct your courses?

MATURANA: It was still in 1973 that I invented a series of lectures entitled *Biology of Cognition*, which later became the book *The Tree of Knowledge*. I gave these lectures year after year, describing the way from the single cell to the social. I was careful not to attack the government in any direct way or to campaign openly for some political end – that was not my thing. I never urged my students to go in a certain direction but I wanted to develop their capacity for reflection step by step.

POERKSEN: If I understand correctly, you wanted to teach them how to think independently. Could you present an example of your teaching to illustrate your approach?

MATURANA: I once talked about my view that power is granted through obedience. Nobody possesses power, I said, but they are given power by others who subject themselves and do what is demanded of them. I had come to the lecture with a very realistic toy gun. "With this gun", I said to my students, "I can kill you." I pointed at a female student and said: "Stand up, or I will shoot you!" She stood up although she knew, of course, that I would never shoot her. – "Come to the centre of the room!" She went to the centre of the room. – "Lie down on the floor!" She lay on the floor. – "Take off your clothes!" At this moment she jumped up and shouted: "No! That I will not do!" I waited for a moment and then said: "You see, this refusal to obey has robbed me of my power. My power rests on your willingness to obey and not on the fact that I am waving a gun about." You see, I did not tell my students what they had to do, but I tried to lead them to other possibilities of reflection and perception. My view is: Those who favour a certain way of living and desire that that way of living arise and reveal itself in the relation to themselves, should live it without hesitation. Waiting will not be of any use.

POERKSEN: Structure-determined systems – human beings – can only be controlled in a limited way; one can perturb them but not control them. Compulsion appears to stand no chance, in principle. My thesis is: You have developed an epistemology that removes the conceptual foundation of dictatorial power.

MATURANA: I strongly support this thesis and want to add that I can destroy the conceptual foundations of dictatorship because my work allows me to achieve a more profound understanding of democracy. Democracy must be created anew every day, I believe, as a space of living together in which participation and cooperation are possible, based on self-respect and the respect of others. The first thing a dictatorship destroys is the self-respect and the autonomy of every single individual, because it demands subjection and obedience as the price for staying alive.

POERKSEN: Could it be that the immense popularity of your ideas today is due to the often-invoked end of all ideologies and the collapse of the sort of socialism that really existed?

MATURANA: I see a connection. What I have written provides a new foundation for the possibility of self-respect, which is fundamentally negated by dictatorships. What the readers of my work may realise is that we are all unavoidably participating in the creation of the world we live in. This is the view that I invite people to try without compulsion or cost, a view that values the individual. And whoever feels appreciated and respected, will be enabled to appreciate and respect themselves. They can accept the responsibility for what they do.

Encounter with Pinochet

POERKSEN: I have been told that you once actually met the dictator Pinochet. Would you like to tell me about the circumstances of this encounter?

MATURANA: One day, it was in 1984, I received a letter with the seal of the president. It was an invitation to have lunch with Pinochet, which had also been sent to other members of the faculty, as I found out later. Some people thought we could not decline, others warned us not to attend the dinner, but I decided to accept the invitation. My mother implored me to remember all the time that I had a family, and I promised her not to forget that. When I finally arrived at the presidential palace, I found that about 85 professors had assembled there. We stood around for a while, talked to each other, and asked ourselves why we had been invited at all. Then Pinochet appeared. His attendant told him the names while he welcomed us. When my turn had come to greet him, I thought of my eldest son who had said to me that he would never shake Pinochet's hand. And there I was, and shook this man's hand. After that, we went to eat in a vast and magnificently decorated hall. As soon as we had sat down, Pinochet rose again, took his wine glass, and said: "Let us drink to our fatherland!" And we rose, drank to each other, sat down again, and ate the delicious meal that was served on elegant porcelain specially manufactured for the President of the Republic.

POERKSEN: You sat there with a man who ran a secret police that spread fear and terror, who was responsible for the disappearance without trace of numerous critics of the government and who ordered people to be tortured.

MATURANA: That is what it was like, precisely. Before dessert was served, Pinochet, who was sitting only a few metres away from me, addressed us again. "Ladies and gentlemen," I heard him say, "the sole purpose of this meeting is to get to know each other. That is all. You may feel quite safe; there will be no demands on you of any kind." He sat down again, and in that moment I picked up my glass, stood up and said: "Ladies and gentlemen, I would also like to toast our fatherland with you!" There was dead silence instantly. One could sense the deep alarm of the assembled persons, who seemed petrified with sudden fear. Pinochet looked at me and leaned forward a little. "We are gathered here today in the company of the president", I went on. "And that is a rare occasion under any government. I will, therefore, seize the opportunity and bring out a toast with you and the president to the effect that we all who are here today contribute to the intellectual freedom and the cultural autonomy of our country, Chile." I drank my wine; Pinochet leant back and clapped his hands four times. All the people in the room clapped four times. One of my friends turned to me and whispered: "Many thanks, that was wonderful." And general talking began again.

POERKSEN: The dictator did not comprehend what you said.

MATURANA: Just a moment, please, the story is not finished yet. Shortly after the dessert was eaten, we all went to another room. A friend of mine, a physicist at our university, pointed out to me that Pinochet was alone and that we should join him. I did not want to at first but he urged me on, and so I finally went with him to join Pinochet who was standing there with one of his generals. "Mister President," my friend said, "I have the pleasure to introduce to you Professor Maturana, a very renowned biologist." I shook his hand again and he said: "I share your good wishes for this country" – "*A dios rogando,*" I answered, "*y con el mazo dando.*" This is a Spanish proverb and means roughly: If you pray to God for something, you must also act accordingly; prayers and pious wishes are not enough. It really was a bizarre situation: Pinochet was standing there and telling me that he shared my desire for intellectual freedom and cultural autonomy. All the goals of his politics were the direct opposite. He wanted to make this country dependent on others in order to be able to crush the first sproutings of communism with the help of his allies.

POERKSEN: You spoke with a man who was thought to be rather limited by many people. Salvador Allende, who had put Pinochet in the position of power in the first place, from which he could venture his putsch, once said that he considered him "too dumb to deceive his own wife".

MATURANA: That was a crass misjudgement. Nobody is made an army general anywhere in the world if he lacks the necessary intelligence. He may be fanatical, narrow-minded and ideological – but he is not stupid.

POERKSEN: What do you think? How did Pinochet understand what you said?

MATURANA: He understood me perfectly well. The essential thing was that I did not treat him as a superior but as an equal Chilean. He was the president for me, he went along with us, and he had to contribute to this grand task of guarding intellectual freedom and cultural autonomy in the country. He was one of us, and that was not meant to be an insult, not at all.

POERKSEN: You re-interpreted the relation between the ruler and his subjects.

MATURANA: One could put it this way – and, furthermore, I used the words he had used in his toast: I also drank to our common fatherland.

POERKSEN: I find this very revealing. You used the eigenlogic of a closed system in order to invade and transform it. You knew, of course, that *fatherland* was an excellent word for that.

MATURANA: Quite so. You cannot, of course, impress an Adolf Hitler with an after-dinner address in which you talk about the Jews and call for their veneration. One must also see clearly that insults cannot be successful in such situations. Whoever does not see and understand that is completely blind.

POERKSEN: This implies, however, that one can exploit the eigenlogic of a system in a subversive way – to put it more generally.

MATURANA: The orientation towards the eigenlogic of the system will work only as long as the meaning or the re-interpretation of what is said cannot be understood as a devaluation of the system. An insult (such as: "You are just a lousy dictator!") would, of course, be quite idiotic because Pinochet would have had to react to it. I was, therefore, extremely careful not to provoke him in any way but to appeal to a common vision: He could not possibly object to a plea for an effort in the service for our beloved country.

Poerksen: How did that encounter end?

Maturana: While we were still talking, another scientist approached and addressed Pinochet in an extremely servile manner. Pinochet stood to attention at once, became the dictator again, and answered brusquely: "What do you want?" I did not want to be associated with this form of servility and withdrew. When Pinochet turned to leave he came my way again, touched my arm and said: "Chao" And I said: "Chao!" I would say that he treated me as a Chilean of equal status because – without being arrogant – I had not subjected myself to him and had not given him power.

Poerksen: Did you ever meet again?

Maturana: No, never. In the evening of the same day, I received two kinds of telephone call: Some people were beside themselves with fury because they thought I had put everyone at risk; others thanked me. One of the professorial colleagues said the wording of my toast had given them back their dignity.

Poerksen: I am quite touched by this experience because it shows that there are always degrees of freedom, behavioural slots, which may be exploited by individuals in different ways. I am sure, however, that such behaviour as yours necessarily depends on talent and intelligence.

Maturana: Such behaviour has nothing to do with intelligence, certainly not. What you need perhaps is a good measure of wisdom based on a capacity of perceiving without prejudices and presumptions. If you approach such a dictator with the image of a terrible idiot and a criminal filling your mind you will inevitably behave in a particular way. Of course, that man is a criminal, no doubt about it. And, of course, he appears completely blind to his responsibility for what happens in Chile and for the horrors of his dictatorial regime,

– as we can tell by his speeches. But if we cling to this assessment we will not be able to see the human being in his prison, with his mental conflicts, and with his patriotism that is, after all, responsibly intended, and to address this human being when talking to him.

POERKSEN: The years of the dictatorship are now definitely gone. In 1989 free elections were held again in Chile; the country now struggles with the problem of an adequate evaluation of its past. If another opportunity should arise to meet Pinochet, who is now an internationally stigmatised old man – although still revered by many Chileans – what would you tell him?

MATURANA: I would advise him to act like Bernardo O'Higgins, the great Chilean freedom fighter. When he was accused publicly one day of having changed into a tyrant, he answered the enraged populace: "Whatever I have done – I have done it with the conviction that it would be beneficial to our country. If the pain and suffering that I may have caused can be relieved by giving my blood then I am prepared to die." Ultimately, O'Higgins was not killed but went into exile in 1823. He was prepared to assume responsibility for his actions and to succumb to the judgment of others. Pinochet has never done that. He still insists that he is innocent. That is his greatest crime.

4. Worlds of science

The paradogma

POERKSEN: During all those years under a dictatorial regime you also worked as a scientist, of course, whose international reputation had started to grow steadily in the early eighties. How did you – quite generally – experience the echo from the world of science? How was your work received? In an essay by Francisco Varela we can read that the first papers you submitted for publication met with total refusal; nobody wanted to print them.

MATURANA: It was not as bad as all that. I sent the first paper directly to Heinz von Foerster and, with his help, it was published in *Biosystems* in 1974. There was certainly a phase of incomprehension but that was no problem for me. On the contrary. When I lectured to the *Biological Society* about autopoiesis for the first time and presented my ideas in detail, a friend of mine came to me afterwards and asked me: "Humberto, what is the matter? Are you ill?" The fact that numerous scientists did not show any immediate interest in what I presented did not, quite honestly, bother me at all. And the critical reviews of my work were not a problem for me at any time because I was always able to demonstrate that the various objections and arguments were not valid. One day, for instance, a colleague said to me that there might be living systems in other parts of the cosmos, which are completely different from the living systems known to us. "How would you know," I asked him, "that we are dealing with living systems if they are completely different? My topic has to do with what is common to everything alive." This is not merely scholastic pedantry but an epistemologically sound argument.

POERKSEN: The current paradigm of normal science is undoubtedly that of realism: the majority of the scientific community still believes in an observer-independent world whose essential features we are able to discover step by step. Such a paradigm often has – to use the words of the philosopher Josef Mitterer – the shape and the hardness of a *paradogma:* The history of science shows many examples of how unwelcome views were classified as non-scientific, how their proponents were marginalised or simply ignored. Have you never worried about such practices of expulsion when they were practised on you occasionally?

MATURANA: No, all this did not concern me because I have never considered myself a revolutionary scientist or a protagonist of some New-Age theory, who has to fight against a particular paradigm of scientific research. I have never longed for recognition or a massive following of fans. I was certainly never disturbed or upset when my work was not properly understood or treated with neglect. This sort of history simply does not apply to me. I was and I am an uncompromising scientist who glided through the years of the dictatorship, always confident and careful to produce impeccable work without logical faults. That is all!

POERKSEN: But were you never irritated by the criticism or the puzzled looks of colleagues and friends? When I appeared the first time in your laboratory in Santiago de Chile about a year ago, something very strange occurred: Whenever you were called to the telephone and we had to interrupt our conversation, one of your colleagues came up to me and said: "You are wasting your time here. What counts is facts. Forget about the observer."

MATURANA (laughing): I know who you are talking about. That is how it is. Some people simply cannot do anything with my views, they find them unacceptable, but they are incapable of refuting them. Occasionally critics tell me that I am really a philosopher, a poet, a mystic. And so on. Such labelling is a way of getting rid of my views and a justification for not having to deal with them any more. Of course, I deeply respect my colleagues but the bad or good opinion they may have of me is of no relevance to me at all. It just does not touch me. When I am criticised or praised then I ask myself: What is the reasoning behind such an assessment? Can I recognise a proper understanding of my ideas in it? Do I share the reasons underlying the criticism or the praise?

POERKSEN: You have just hinted that people keep worrying whether you are better appreciated as a philosopher or as a scientist. This insecurity concerning the classification of your ideas comes out clearly in a short anecdote: On the notice boards of your institute one could read for many years *Instituto de Neurobiología*, then *Epistemología Experimental*, and finally the hybrid expression *Neurofilosofía* appeared. My question is now: How would you describe yourself?

MATURANA: Perhaps I could be characterised best as a humanist philosopher who – equipped with the knowledge of modern times – has reverted to the time before the separation of science and philosophy. When Galileo separated philosophy from science, he separated, as I would say, theories that contained and preserved different things. The purpose of philosophical theories is to preserve principles. Experiences, which do not contribute anything to the preservation of these principles, are judged irrelevant; they are abandoned and remain neglected. The purpose of scientific theories is, however, to maintain coherences with experience. Principles can, therefore, be liquefied – and in this way a scientific theory arises. Naturally, Galileo did not describe such a distinction in these words, but with the actual completion of his division the philosophers, who had devoted their efforts to principle-directed reflection, lost all contact with the world of experience. In my work I am re-uniting philosophical reflection – that is, the analysis of the foundations of what one does – with science and scientific theory formation.

BETWEEN PHILOSOPHY AND SCIENCE

POERKSEN: How did you arrive at this somewhat extraordinary distinction between philosophy and science?

MATURANA: It goes back to an experience in Bregenz. Philosophers and adherents of Karl Popper had invited me to criticise the evolutionary epistemology developed by Konrad Lorenz. I did not wish to do that because I am not interested in criticising such an outstanding biologist like Lorenz, although we definitely hold very different views. In my lecture, I therefore dealt with the closure of the nervous system and tried to show quite generally and with reference to any kind of epistemology that no human being can gain access to an independently existing reality. In the following discussion, the problem of reality dominated everything. Somebody stood up and asked me: "Have you published anything?" – "Of course", I replied, "You may find articles of mine in various journals kept by your library." – "Will I find there", he wanted to know, "your *real* articles?" Things continued in this way. At the end a philosopher joined in and said: "Having listened to your lecture, I am full of admiration. Never before have I met a person who could use the English language in such a beautiful way in order to say absolutely nothing."

POERKSEN: That really does not sound like a compliment.

MATURANA: Sure. I asked myself, consequently, what all these renowned and undoubtedly knowledgeable and cultured people who had come together there, actually wanted to tell me. Finally, the thought took shape that there is a fundamental difference between philosophical and scientific theories: the people designing and formulating them want to preserve different things. I can only repeat: If the goal is to maintain coherence with what can be experienced, then one generates scientific theories. If one wants to preserve principles, then one generates philosophical theories: Experiences that do not fit the principles are abandoned, discarded, and devalued. In this respect, a philosophical theory shares strong similarities with an ideology: What must be preserved unconditionally from the perspective of these philosophers is the idea of an observer-independent reality that should remain uncompromised. And that is why they were dogmatically bound to ask questions of one orientation only.

POERKSEN: Could you perhaps specify more precisely the special mixture of philosophy and science that we find in your work? Could one say that you pose philosophical questions and give scientific answers?

MATURANA: Practising philosophy means, I claim, reflecting about the foundations of what one does. That is certainly what I do, and therefore one may call me with some justification a philosophical thinker. Searching for an answer, however, I proceed as a scientist; I use experience as my orientation and design scientific theories. What you find in my work is indeed a mixture of philosophical questions and scientific answers – this seems to me a correct observation. But if the problem is to select a fitting label, I would prefer to call myself a biologist who is trying very hard to keep two different domains separate: the domain of the internal dynamics of a system and the domain of the interactions of that system.

POERKSEN: In your books, you practically never refer to philosophical precursors. Are there none? Have you developed your *neurosophy* without paying attention to its underlying traditions?

MATURANA: Naturally, I have read some philosophers. I occupied myself with Plato, for instance, and his wonderful conception of the ur-idea, but his approach was without any relevance to my work as a biologist who was interested in the structure of a living system and the processes resulting from that structure. I found Hegel's *Phenomenology of the Spirit* and its description of master and servant fascinating, but none of my own insights come from there. Reading Nietzsche's *Thus Spake Zarathustra* was also extremely illuminating for me, but there was no reason for quoting it as one of my sources. I read a bit of Kant, studied some Heidegger and Sartre, and took an interest in Merleau-Ponty. The questions of interest to me do not, however, result from this reading of mine because all these authors – even when they speak of biology – argue as philosophers, and that means that they are always bent on preserving their principles when generating theories. They are not biologists, and I am not a professional philosopher.

POERKSEN: But I think that we might be right in saying that you are arguing as a scientist and reaching the same conclusions as philosophical epistemology? It has been noted variously, for example, that your ideas correspond with those of Kant, who focusses on the *transcendental subject* – and shows that all perception is unavoidably structured and that the absolute, the thing-in-itself, is unknowable.

You are concerned with the investigation of the *empirical subject* – and describe the observer-dependence of all knowledge. The conclusions are similar.

MATURANA: The similarities you may discern in the conclusions do not indicate any deeper-reaching agreement. Here is a modest analogy: Imagine two curves intersecting at a certain point. The coordinates of this point of intersection are the same for both curves, but each of the curves has a different inclination and a different trajectory. Although Kant and I occasionally seem to reach similar conclusions, we make fundamentally different statements and proceed from different backgrounds. Kant takes the path of philosophical reflection, I argue as a biologist. He speaks of the unknowability of a thing-in-itself, of an absolute, independently given, reality that is for him the ultimate point of reference. I maintain, however, that it is meaningless to speak of a thing-in-itself, even when conceding that it is unknowable. The existence of this thing-in-itself cannot be validated in any way because all that we can say about it is dependent on our personalities and our perceptions.

NOTES OF AN OBSERVER

POERKSEN: An observer arranging your theories by chronological and publication dates may establish four different stages in your development. At the beginning, you practise as an empirical biologist, investigate frogs, pigeons and salamanders in your laboratory, and publish your results in the domain of *neuroanatomy*. Then you develop a *bio-epistemology* concerned with the question how a living being creates and produces its own world. It is followed by your critique of the ideal of objectivity and of truth fanaticism, i.e. a stage of *bio-ethics*. You expound how the, biologically indefensible, belief in the possession of absolute truth leads to the oppression of people thinking along different lines. Stage four, finally, deals with the general foundations of being human, with a sort of *bio-anthropology*. Here now the focus is on love as the basis and foundation of human co-existence. What do you think? Does this kind of categorisation do justice to your development?

MATURANA: Listening to what you are saying actually makes me recognise these different stages in my work although such a genetic ordering has never had any impact on me. It does not correspond to my own experience. I would rather say that I have always been carrying around a whole set of fundamental questions. For example, I already wanted to understand life and death when I was only a child. These fundamental questions were my constant companions when I was a student and when I worked in the laboratory, and they inspired me to search for ever more thorough reflection. I always try to discover what the reasons are that lead to an assumption. What processes constitute a particular entity? How do I know that I have found the right answer to one of my questions? Why does a certain view appeal to me, and not another?

POERKSEN: You literally acquired widespread fame in the mid-eighties, having been known before primarily among biologists and cyberneticians. *Autopoiesis* suddenly became a universally fashionable term, and sociologists, management consultants, and psychotherapists all over the world picked up your ideas. Your impressive popularity has always been a cause of wonder to me because you are actually a difficult thinker. The language you use is not at all easy to understand; you re-interpret many concepts, you invent new words, and you expect a lot from your readers – in brief: you are not really aiming at the big audiences.

MATURANA: I do not believe that my considerations are particularly difficult to understand but rather that they may be particularly difficult to accept. It is also not the case that I have invented so very many new concepts; I have always taken great pains to use concepts with a strictly defined meaning and to avoid metaphors because they may impede and even prevent understanding what I mean. That is to say: The problem of comprehensibility is, in my view, actually a problem of acceptability. In most cases people seem not to understand what they do not like and do not want to hear or read. Therefore, they raise questions, hoping that what they heard and, in effect, understood but did not agree with, might in a repetition turn out to be different, after all, from what they understood and for some reason rejected.

POERKSEN: You write in a decidedly abstract style, cutting out unconventional metaphors, parables and personal stories. But does not abstraction contribute to rendering the observer invisible? Abstraction detaches a thesis from the particular experience it may possibly be derived from.

MATURANA: I do not agree. Naturally, I use abstractions in my writing but they are derived from the coherences of what we may experience; they are, therefore, comprehensible and will stimulate readers who want to know more. The alternative of using stories, metaphors, and images does not appear meaningful to me at all. I do not consider it a good idea to present the observer Humberto Maturana and his personal experiences; and I do not want to do this because we are not dealing with the operations of a particular observer but with the operation of observing in general. The crucial insight is that observers specify what they perceive by their operations of distinction – that is what matters. I do not use metaphors, either, because they mix domains: they appear to be easy to understand; in effect, however, they damage the understanding. I believe that metaphors lead us astray – and I have therefore introduced the term *isopher* for statements that themselves exemplify what is just being discussed or described, statements that are exemplars of what they are expressing. Here we do not connect or mix different domains, as in the case of metaphor, in order to achieve understanding.

THE DOORS OF PERCEPTION

POERKSEN: How did you experience that surge in popularity in the world of science? For a period of time you were compared with either Immanuel Kant or Ludwig Wittgenstein, *Biology of Cognition* was classified as the most important article of the century, and its author as a "rising star." The reverence shown towards your person was sometimes a bit weird. A famous cybernetician and early protagonist of systemic and ecological thinking is reported to have uttered on his death bed that the essential impulses for an understanding of the living world can be expected to come from a certain Humberto Maturana in Santiago de Chile.

Fig. 15: The Crown of Thorns *by Hieronymus Bosch*

MATURANA: Obviously, my ordinary life changed a little due to the euphoria with which my work was received. There were innumerable invitations. I was once called the *Edith Piaf of neurophysiology*. My popularity allowed me to travel more, to meet many people, and to earn some money. Basically, however, I think that I was merely a passing star in many domains. First people applauded my thesis in neuroanatomy, then my work in neurophysiology, finally the paper *Biology of Cognition*. And so on. Then, one day, a different topic appeared to be new and central; the enthusiasm of individuals is always of limited duration. Things pass. I have never put great store by the compliments that were made to me. I listened to them, expressed my thanks – and let pass what was said. This should not be misconstrued as arrogance – but I am quite aware of certain temptations, in particular the temptation of fame. Do you remember the painting by Hieronymus Bosch, which is reproduced at the beginning of the book *The Tree of Knowledge*?

POERKSEN: We see Jesus surrounded by some people.

MATURANA: I shall leave it to you to decide whether you want to include what I am now going to say in our book; I trust your good judgement. In 1962, I received a call from a friend of mine who was studying the mind-extending effects of psychedelic drugs. Many people, at the time, were influenced by Aldous Huxley's essay *The doors of perception*. This friend kept inviting me to LSD experiments but I declined because I did not have any questions that I wanted answered under such circumstances. One day – it was in 1963 – he called again, and this time I accepted because a relevant question had emerged. I wanted to know whether the nervous system of a human being continues functioning normally after taking LSD. We met one evening in my house. The children were already in bed. We listened to music. There were a few books on the table. The LSD I was given by my friend came in the form of small impregnated pieces of paper with different pictures on them. These pictures –the jaguar, the sun, and the moon – indicated a particular quantity. I ate the sun, and my eyes fell on a book and the painting by Hieronymus Bosch, the *Crown of Thorns*. I contemplated this painting for several hours. What did all these different people want to tell Jesus, I asked myself. Finally, the thought struck me that the four people stood for four different temptations. Of course, this interpretation is entirely my own. The old man caressing Jesus' hand embodies the temptation of indifferent superficiality. He seems to be saying to Jesus, who appears involved and concerned but calm and collected: "Keep away from everything, and you will live to a very old age!" Another man is apparently whispering something in his ear, trying to create the impression that he has something to tell; he represents the temptation of vanity pretending to be modesty. The man putting the crown of thorns on Jesus embodies the temptation of envy. He seems to be dissatisfied with himself and devalues himself in the comparison with someone else. The fourth figure in the painting has grabbed Jesus' cloak and tugs at it, restraining his freedom and his possibilities. For a very long time I was unable to understand the meaning of this figure. Many years later, I hit upon the idea that the man represented the temptation of certainty: He lives in a world without alternatives, in a world without reflection.

Poerksen: How do you relate these four temptations to the question of the prospects and the dangers of fame?

Maturana: I would say that the temptations of vanity, of superficiality, envy, and certainty are immediately active when you are better known and suddenly find yourself admired by people. Perhaps you start to believe in the catalogue of flattering attributes and behave accordingly. It is a form of captivity to be considered someone special. Moreover, those who identify the attributions of others as their outstanding properties seem to be blind for me: Whatever someone else sees in me – it is never my own self, it is never my own personality.

IV. Ethics of a theory

1. The biology of love

THE TWO IDENTITIES OF THE SCIENTIST

POERKSEN: At the end of your monograph *Biology of Cognition* you point out that a scientist must be aware of, and face, the consequences of his work – in brief: its ethical or unethical consequences. This means: science, for you, is not a value-neutral activity.

MATURANA: Naturally, many scientists believe that they are neutral and objective and have nothing to do with the object of their research. I do not share this view. Science is not a domain of objective knowledge but a domain of subject-dependent knowledge, defined and determined by a methodology, which lays down the properties of the knower. Pure science does not speak to us; scientists speak to us, who bear the responsibility for their pronouncements. Scientists do not describe an objectively given world, a transcendental reality. They can grasp only what they are able to distinguish and investigate. They describe what they hold to be relevant, what they want to observe, show, and prove experimentally in certain ways.

POERKSEN: What follows from this kind of insight? Better: What should follow?

MATURANA: The scientists who are aware that whatever is said is said by them, also know that their research is bound to have consequences for other people. They must, therefore, make clear the connections between their work and the ethics of the world they live in. They must, in effect, live with two identities: on the one hand, the identity of the scientist whose task it is to explain experiences by proposing generative mechanisms, on the other, the identity of human beings who reflect the consequences of their actions.

POERKSEN: Many scientists refer to social responsibility, when they discuss ethical questions. The essential and constructive focus of your own ethical reflections, however, is on another concept, which seems unusual in this context and has emerged only sporadically in our conversations so far: *love*. How do you connect ethics and love? What is love?

MATURANA: Whenever we see a relational behaviour through which another arises as a legitimate other in the domain of coexistence in which it takes place, we speak of love. As such love is a manner of relating, a relational domain, that occurs spontaneously practically in all living systems, particularly in mammals and human beings. Love is the fundamental relational domain in which human beings exist, and constituted the relational conditions for our evolutionary origin. We feel well when we can look after others. I claim: Love is a feature of human co-existence. It opens up the possibility of reflection and is based on a form of perception that allows the other to appear legitimate. In this way, a space arises in which cooperation seems possible and our loneliness is transcended: The other is given a presence to which we relate with respect.

POERKSEN: Understanding the concept in this way seems a bit difficult. Speaking about love in ordinary language usually conjures up images of twosome harmony: strolling along the beach together, kissing, embracing. You are not talking about that.

MATURANA: Not necessarily. Naturally, we will embrace someone when we feel there is a common longing for an embrace. I do not refer to this kind of loving intimacy when I am speaking of love. Perhaps an example will help: You are strolling along the beach and notice suddenly that a child is washed into the sea by a wave. When you then rush into the water and save the child from drowning, you are acting out of love. However, when you summon the child and give it a dressing down, then that is not loving activity: you do not pay attention to its fears but obey only your own anxieties. The emotion ruling your activities in that moment is your own fear. A behaviour based on an adequate perception of the child would be stroking the child in order to reduce its fright and to show him or her how to move around the beach without risk.

POERKSEN: How far does this loving acceptance extend that you are describing? Does it include the relationship between humans and animals?

MATURANA: There are many examples that show modes of behaviour here that we would call love. It is obvious in the case of the dog jumping around you wagging its tail and being stroked in return. However, there are also less obvious cases of love between humans and animals. Let me present a little story about what I once experienced in Bolivia. We were sitting together after supper, an agreeable group of people in pleasant spirits, smoking and talking. Suddenly a spider came down in the middle of the table. One of the guests excitedly reported the visit of the animal to the hostess: "Look here, there is a spider!" – "Nothing to worry about," the hostess said, "it always comes round after supper to collect the leftovers, and then climbs back to its hiding place." I insist that that woman and the spider lived in a social relationship; each had legitimate presence. The spider was left alone, and appeared only when it did not disturb people eating. What we observed was love.

TRUSTING EXISTENCE

POERKSEN: You once said that 99 percent of all illnesses were caused by a lack of love. You added by way of qualification that you might be wrong: it could be only 97 percent but definitely not less. How are we to understand that? What connection do you see between a lack of love and illness?

MATURANA: The fundamental condition of existence is trust. When a butterfly has slipped out of its cocoon, its wings and antennae, its trunk and its whole bodyhood trust that there will be air and supporting winds, and flowers from which to suck nectar. The structural correspondence between the butterfly and its world is an expression of implicit trust. When a seed gets wet and begins to germinate, it does so trusting that all the necessary nutrients will be there for it to be able to grow. When a baby is born, it is completely trustful that there will be a mother and a father to take care of its well-being. This implicit trust, however, on which the existence of all living beings is based, is disappointed continually. The flowers are poisoned by insecticides, the seeds lack water, and the baby, appearing in this world as a loving being, does not receive love, is not seen but negated in its existence. I maintain that the continual negation of the other produces illness, that is to say, the loss of organic harmony both inside the organisms and in the relation with their existential circumstances. The systemic dynamics of a human being will, when it is negated permanently, change in such a way as to destroy its original harmony and induce further destructive challenges and stresses to the body that will in turn cause additional disharmonies. The results are increasing susceptibilities to infections and somatic and mental illnesses.

POERKSEN: Could we use your description of love to reveal the ways and manners of humans living together? Your understanding of love could then serve as an instrument and a stimulus of knowledge, a foil of contrast to precise description.

MATURANA: But of course. Once it is understood what love is, it can immediately be ascertained when and under what conditions love is negated. We can observe the parents who permanently correct their children, reprimand them for their mistakes, and threaten them with punishment. We can perceive the characteristics of our culture and realise that the highly praised idea of uninhibited competition is not a source of progress but generates blindness and restricts the opportunities of co-existence because it negates the other. Ambition, mistrust, the culturally anchored pursuit of power, and the passion for control are the forces, we become aware, that cause love to disappear. The economisation of relations – claims are *exchanged*, needs *negotiated*, compromises *enforced* – destroys the pleasures of simple togetherness because it is organised according to the patterns of commercial business practice. The basis of a partnership is no longer mutual trust and reciprocal respect but the singleminded negotiation of advantages.

POERKSEN: What happens when the other is no longer seen? Can you find an example for this technology of negation?

MATURANA: In the early sixties, when the Americans began to get involved in Vietnam, I discovered the headline in the *European Times*: "50 Americans murdered! 200 commies exterminated!" Here a decisive difference manifests itself: For the author of the headline, the Americans had legitimate presence, but not the "commies." Their destiny did not matter; they were not murdered but simply "exterminated." This also implies: Ethical involvement does not extend beyond one's own sphere of social belonging.

POERKSEN: In the light of this headline, the other no longer appears to be a human being with whom one has anything in common.

MATURANA: Indeed. One possibility of destroying ethical impulses in the parties involved in a war is to deny the opponent the features of human beings: The enemy is dehumanised, downgraded to an "untermensch" and "extremist", a "communist" or "Nazi". The guidelines for soldiers in a war include that they kill first and reflect afterwards. Only those who are willing to extend the domain of legitimacy of the other, the domain of love, in such a way as to cover all human beings, and who refuse to be driven by discriminating designations, can be moved by the fate of each and every individual and include them in their ethical reflections.

POERKSEN: In what ways may we live together if our actions are guided by love?

MATURANA: It is possible to talk things through, to discuss and study problems together, and to participate in the accomplishment of common tasks that are relevant to different people. Nobody has to apologise for their existence and their experiences but people exist in a domain of cooperation, which has the characteristics of a social domain. In more general terms: We see here a democracy because love is the emotion constituting democracy. The fundamental features of democracy include human beings – citizens – who respect themselves and each other living together and working together on a project and a form of co-existence. Is it not revealing and extraordinary that there are no citizens in a monarchy or a dictatorship? Here you are compelled to obey and subject yourself unconditionally, however friendly or prudent the king of a country or a tyrant may show themselves, here you are a subject or a slave but not a citizen.

POERKSEN: Would you say that a form of living together that is founded on love is more stable than a dictatorship? We have experienced often enough that tyrannical rulers may cause horrific devastations but cannot maintain their domination permanently. Adolf Hitler's 'Reich' of a projected 1000 years lasted for just about 12 years.

MATURANA: It need not always work like that because a system will exist as long as the conditions for its constitution are maintained. A perfect dictatorship systematically eliminates dissidents and so averts its collapse. However, if people under such circumstances discover love, they will stand up and rebel against their constant oppression and continual negation as individuals. A dictatorship projected for a time span of 1000 years would ultimately have to transform the whole world into its own system and kill all those who do not conform and rise up against it. It requires enormous efforts and massive use of force to maintain such domination; you need a police force, bodyguards, and instruments of manipulation – but a dictatorship that remains stable for a long time is not impossible. However, if only one individual survives that succeeds in preserving and teaching the ideas of love and mutual respect for others, then resistance will develop again. Love generates such well-being and is so liberating that many people risk their lives in order to spread it and defend it.

POERKSEN: What are the consequences of this thought? Do they converge in the old hippie slogan: *Make love, not war*?

MATURANA: No. We human beings attribute different values to different emotions, and we, therefore, occasionally suppress the realisation of these emotions. Commandments of all kinds always tend to manoeuvre us fatefully close to the roles of missionaries and tyrants. They are suitable instruments of discrimination: "We are all for love here", people may then pronounce with an air of superiority, "but the others go to war!" I therefore do not preach love, I do not formulate any commandments, I do not, in fact, recommend anything, neither love nor indifference, neither friendliness nor hate, but I say: if there is no love, there are no social phenomena, no social relations, and no social life. The emotion constituting social life is not hate, not self-interest and greed, not competition and aggression, but love.

POERKSEN: But it is evident, no doubt, that human social life is not only shaped by love.

MATURANA: Of course, anger, hatred, envy, and various other emotions shape the actions and relations of our social life. Of course, there are diverse variants of communal existence that are not based on love. Just think of a monarchy, of some ideologically or religiously determined sect, or of an army; their constitutive hierarchies always lead to the disappearance of individuals. My claim is that there are no social relations in an army – if we disregard the friendly personal relations between soldiers or generals. Sometimes small islands of social relations may form within such wholes that are organised along different lines. Social life, however, I insist, is based on love.

SOCIAL SYSTEMS

POERKSEN: Can you see no contradiction between the individual and the social? Those who speak of the individual and emphasise individuality presuppose, as a rule, that the individual is autonomous, a monad, and inaccessible to external impressions. Those who, in contrast, foreground the formative power of the social, usually insist on the permeability of the individual: Individuals, the assumption is, observe the world through the eyes of their group and against the backdrop of their history. Both views clearly contradict each other.

MATURANA: This is not my view. I think there is no contradiction between the individual and the social: a society is a collection of individuals living together on the basis of a fundamental emotion. The member of a social community is and remains unavoidably an individual. When individuals speak with each other, make appointments and do things together, they surely do not lose their individuality; they may change their views and may be transformed through their encounters but they continue to exist as individuals in their autonomous dynamics. In their interactions they create something new that cannot simply be attributed or even reduced to merely one of the involved persons. Should their individuality actually be diminished or lost completely, e.g. due to an illness, they will no longer be fully responsible members of a social community. In an army, on the other hand, which is undoubtedly not a social system, individuals are clearly unwelcome. An army can only use agents, executors of planned actions, people who carry out orders without reflection. Persons who cannot conform to army rules will be thrown out.

POERKSEN: You are one of the few sociologically interested scientists who do not use biology to devalue the individual. In the history of social Darwinism, we can find many contrary positions: biology supplies the arguments supporting the dominance of collectives and the degradation of individuals.

MATURANA: Such patterns of argumentation and procedures of justification are, however, not grounded in an adequate understanding of biological processes. These explanations and ideas are intended to serve particular purposes and are actually invented: they are then projected onto biology and nature and, in a second step, referred back to the human domain in order to lend support to the original presuppositions. Charles Darwin borrowed the idea of competition from the economists of his time in England. Some time later, the economists adopted the idea of competition from biology in order to validate their own economic paradigms. – Let us assume that we would very much like to create a theory of society proving the dispensability of the individual and the overarching importance of the collective. We would invent a frame of reference in which the collective is given the highest value. At the same time, we would have to turn a blind eye on the fact that the components of collectives are undeniably individuals whose autonomous dynamics are preserved in the interactions with others. And only as far as they are and remain individuals, and therefore contribute to the maintenance and the progress of the well-being of the collective, can we speak of a social system at all – and not, for example, of an army, a monarchy, or a dictatorship. Therefore, I claim that individuals are not dispensable.

POERKSEN: A point of conceptual clarification: What form of living together do you actually designate as a *social system?* The term is generally used in a comprehensive sense to refer to the entire set of structures made up of human relations.

MATURANA: If you listen carefully to the characterisation of a certain kind of behaviour as unsocial you will notice that it revolves around the lack of respect for the other. We complain about people's unsocial behaviour when they act without respect, when they, for instance, simply throw their rubbish across the garden fence on the neighbour's property. The complaints we usually hear in such cases always indicate an emotion. With this conceptual specification I definitely do not intend to give a definition of what is social, I just consider the conditions that lead us to describe certain behaviours in our everyday life as unsocial or as social. One of the customary self-defining characteristics of sociology is that all human relations are social relations – this is a view I do not agree with at all. What gives a human relation its particular character is its special emotional foundation. Once this is understood it is no longer difficult to grasp that all those relations, which we call social relations in our ordinary language, are based on love.

POERKSEN: If social systems can only be categorised as such when they satisfy particular demands – acceptance and appreciation of the other –, the question arises as to what the actual tasks of sociologists, the professional observers of society, might be. What are their topics? What forms of living together remain legitimate objects of sociological analysis?

MATURANA: Sociologists should deal with the emotions underlying human relations. Their task should be to show how these emotions mould and style the ways and modes of communal life. I once suggested distinguishing *homo sapiens amans* from *homo sapiens aggressans*, and the latter from *homo sapiens arrogans*. All these concepts have to do with fundamental emotions like love, aggression and arrogance, which have influenced the patterns in which relations were established during the course of human evolution and formed the existence of *homo sapiens sapiens* – the human being living in language.

POERKSEN: You seem to consider emotions, and not rational arguments, as the essential determining forces.

MATURANA: Emotions guide us. People, who transform their relationship with other persons in an all-encompassing way, realise on closer inspection that they have, in effect, fundamentally changed their underlying emotion. Emotions are, in my view, dispositions for action; they seem to me to be something completely elementary that also determines the acceptance or rejection of a rational system. All rational systems and discussions rest on a foundation that is non-rational in kind, and that is accepted owing to personal predilections. It may, therefore, easily happen that we rationalise our actions resulting from such predilections so as to justify them after the event: rationality is then plainly a method and an instrument of justification. For me, humans are emotional animals that use their minds and their rationality for the purpose of denying or justifying emotions.

POERKSEN: Such a description makes me feel somewhat uncomfortable. You might interpret this feeling of discomfort as the typical prejudice of a representative of the arts faculty. Anyway: Does not your characterisation amount to a devaluation of human beings as rational animals?

MATURANA: Not at all. It is a distinctive feature of our culture to devalue emotions as forces interfering with, and even threatening, rationality – here you have actual devaluation. What I am saying is, however, that love is the prime emotion that makes ethical behaviour possible, in the first place, a behaviour, which includes the responsible reflection of the consequences of actions. Ethical concern arises at the moment when self-awareness emerges and when, therefore, the possible consequences of one's actions for another human being of personal importance are consciously reflected. Ethics is, for me, a consequence of love; it occurs in language because only language enables us to reflect our chosen course of action.

Ethics without Morality

POERKSEN: What happens when conflicts arise? Can there be no rationally controlled solution?

MATURANA: Any successful resolution of a conflict is of an emotional nature. This does not imply that I am pleading for stopping all discussion and giving up all conversation, not at all. What must be achieved is the creation of a common basis that permits reconciliation and relieves the parties in the conflict of their fears. When people speak with each other in order to resolve their conflicts, they must begin by first restoring mutual trust and respect. It may perhaps be advisable to admit mistakes, to apologise and to affirm the intelligence of the other person. When reciprocal trust has been restored, people will listen to each other in a way that grants validity to what is said in the relevant domain of reality. On such a basis a new common emotional dynamics may be developed, which can sustain the relationship. The old certainties are abandoned and a sort of behaviour is resumed that I call love.

POERKSEN: It seems to me that your reflections dealing with love and the power of emotions always involve an unwarranted jump: there is a leap from the facts and arguments of hard science to poetic descriptions applied to practical fields, from the characterisation of what there is to what should be, from epistemology to ethics. You are changing the discourse.

MATURANA: This is incorrect. Biology does not tell us what we must do, and as a biologist and therefore as a scientist I cannot tell anyone what to do – that would be a misunderstanding. In nature, nothing is good or bad. Things simply *are*. It is only in the human domain of the justification and rejection of a certain kind of behaviour – i.e. when our particular preferences are at stake – that evaluations and distinctions such as good or bad arise. Once again: I do not give any recommendation but I can, for instance, assert as a biologist that tampering with the genome will produce monsters. But that does not at all entail that I am arguing for or against the manipulation of the genome, it only means that I describe the consequences resulting from a particular course of action. People may then make their choice.

POERKSEN: Does not the specific mode of description contain a partisan view and an indirect plea?

MATURANA: No. Perhaps people's reception is influenced by their own values and preferences – but that is another matter. In such cases, it is obviously difficult to perceive simply what is happening and what is shown.

POERKSEN: But is the concept of love not already connected with a positive valuation? The word *love* sounds so good. Nobody in his right mind will openly promote exploitation and dictatorship.

MATURANA: If I want to keep valuation and description clearly separate, all I have to do is to argue as clearly and as precisely as possible and state exactly what I mean and what I want to say. Naturally, whenever I observe a behaviour that involves another person as a legitimate partner, I could use *Num*, – a new and unencumbered word. People would then probably ask me why I was using such an expression since as the word *love* was the common concept available for this kind of behaviour and this trace in the flow of relations. I would like to repeat once more: I have no intention whatever of promoting love but I do indeed insist that there can be no social phenomena without love.

POERKSEN: Nevertheless the idea suggests deriving an ethical imperative from your considerations. We might say: *Act always in such a way as to create or preserve love.*

MATURANA: We could say that, of course, but the formulation of an imperative turns ethics into morality. I would like to propose at this point of our conversation that we distinguish clearly between ethics and morality, even though such a distinction may at first sight appear somewhat arbitrary. The moralists stand for the adherence to rules, which they consider as the external reference lending authority to their statements and strange ideas. They lack awareness of their own responsibility. People acting as moralists do not see their fellow human beings because they are completely occupied by the upholding of rules and imperatives. They know with certainty what has to be done and how everybody else has to behave. People acting ethically, on the contrary, perceive others, consider them important, and see them. It is, of course, possible that persons argue like moralists but act in an ethical way. It is imaginable that persons are moralists without being ethical, or that they are generally held to be immoral while, in fact, acting ethically. In each of these cases, the possibility of ethics and of being touched arises only when the other human being is seen as a legitimate other, and when the possible consequences of one's actions for that other's well-being are reflected. Ethics is based on love.

POERKSEN: What do you say to those who – despite your determined refusal to formulate rules and imperatives – see some similarity here with the Christian commandment that we should love our neighbour as we love ourselves?

MATURANA: Jesus spoke of the love of the neighbour. The Christian churches, however, that have been involved in wars and devastations, have for 2000 years interpreted what Jesus said as a commandment. We might rather say that we should always have a gun ready and the finger on the trigger if we cannot trust our neighbour. The question is then: Do we really want that? Do we want to run around with a gun day and night and live in a world ruled by anxiety and mistrust? If this is what people want, then they must not love their neighbours and not trust them in any way because the neighbours will then be justified in mistrusting and fearing them as well. In this way, an apparent reason is created for carrying arms. Conversely: If you behave in a way that grants other people respect, you will in turn earn their respect. If you trust a child, the child will in turn trust you. This does not mean that I am now supporting the view that what we do not want to experience and suffer ourselves we should not do to others; that would only be opportunism, not love. All I am saying is: *We bring forth the world we live by living it.* Whatever we wish we should do.